O level Additional Mathematics

O level
Additional Mathematics

M. BROWNE

B.Sc., A.K.C.
Head of the Mathematics Department
Gorleston Grammar School

Methuen Educational Ltd

LONDON · TORONTO · SYDNEY · WELLINGTON

First published in 1962
Reprinted three times
Second (*metric*) *edition* 1970
Reprinted twice
Reprinted 1974
© 1962, 1970 *by M. Browne*
Printed Offset Litho in Great Britain by
Cox & Wyman Ltd,
Fakenham, Norfolk
I.S.B.N. 0 423 80280 1

Contents

Preface

This book has been written for pupils who have reached the 'O' level of Elementary Mathematics, especially those in the fifth form. It has been found suitable for pupils who are taking Additional Mathematics after using it for one year, and as a foundation for the separate Mathematics subjects at 'A' level.

It covers all the topics in the Cambridge Additional Mathematics Syllabus and the majority of the topics in the syllabuses of the other examining bodies. As this is an introduction to advanced mathematics, demonstration and reasonable assumptions have been used where rigid proofs are not necessary. Sections 7.2 to 7.6 are treated as revision. The addition formulae of section 14.1 are most easily proved by using the rotation matrix. The mechanics chapters are based on the SI unit of force, the newton. In section 28.3 the definition of the quartiles takes the median as inclusive.

In the Pure section the high cost of revision has prevented the removal of the full stop after the abbreviation cm, and the change from $\log_{10}$ to lg. The diagrams have had to be reduced in size during printing and the reader is asked to draw the scale diagrams to the original scale. The sides of the graph squares are one centimetre.

I should like to thank all who have helped in the preparation of this book.

M. BROWNE

Caister-on-Sea
January 1970

PART ONE

Pure Mathematics

1. The Binomial Theorem
for Positive Integral Powers

1.1

We can work out $(1+x)^2$ by inspection to give $1+2x+x^2$. When $(1+x)^3$ is required the previous result is multiplied separately by 1 and then x, and the two results added to give $1+3x+3x^2+x^3$. It would be very tedious to do this for higher powers of the binomial $(1+x)$ and so Newton's binomial theorem is used. We allow n to represent the power, giving the expansion:

$$(1+x)^n = 1+\frac{n}{1}.x+\frac{n(n-1)}{1.2}.x^2+\frac{n(n-1)(n-2)}{1.2.3}.x^3+$$

$$+\frac{n(n-1)(n-2)(n-3)}{1.2.3.4}.x^4 \ldots x^n$$

Applying this to $(1+x)^7$ we obtain:

$$(1+x)^7 = 1+\frac{7}{1}.x+\frac{7.\overset{3}{\cancel{6}}}{1.2}.x^2+\frac{7.6.5}{1.2.3}.x^3+\frac{7.6.5.\cancel{4}}{1.2.3.\cancel{4}}.x^4+$$

$$+\frac{7.\cancel{6}.\cancel{5}.\cancel{4}.\overset{3}{\cancel{3}}}{1.2.3.\cancel{4}.\cancel{5}}.x^5+\frac{7.\cancel{6}.\cancel{5}.\cancel{4}.\cancel{3}.\cancel{2}}{1.2.3.\cancel{4}.\cancel{5}.\cancel{6}}.x^6+\frac{7.\cancel{6}.\cancel{5}.\cancel{4}.\cancel{3}.\cancel{2}.\cancel{1}}{1.2.3.\cancel{4}.\cancel{5}.\cancel{6}.\cancel{7}}.x^7$$

$$= 1+7x+21x^2+35x^3+35x^4+21x^5+7x^6+x^7$$

Exercise 1a. Expand by the binomial theorem:

1. $(1+x)^4$ 3. $(1+x)^5$

2. $(1+x)^6$ 4. $(1+x)^8$

1.2

Before Newton formulated his theorem the binomial coefficients were found from Pascal's triangle, which is built up by starting a line with one and then adding pairs of numbers in the previous line:

$$\begin{array}{ccccccccc} & & & & & 1 & & & \\ (1+x)^1 & & & & 1 & & 1 & & \\ (1+x)^2 & & & & 1 & 2 & 1 & & \\ (1+x)^3 & & & 1 & 3 & & 3 & 1 & \\ (1+x)^4 & & 1 & 4 & & 6 & & 4 & 1 \\ (1+x)^5 & 1 & 5 & & 10 & & 10 & 5 & 1 \\ (1+x)^6 & 1 & 6 & 15 & & 20 & & 15 & 6 & 1 \\ (1+x)^7 & 1 & 7 & 21 & 35 & & 35 & 21 & 7 & 1 \end{array}$$

Thus the final line was built up

$$(1+6) = 7; (6+15) = 21; (15+20) = 35, \text{ etc.}$$

This triangle forms a useful check and also shows the symmetry of the binomial coefficients. It should also be noticed that there is one more term than the power n. The odd powers will have an even number of terms, the middle two having the same value. The even powers will have an odd number of terms, giving an odd term in the middle. The full expansion is given by putting in the powers of x, getting one larger with each coefficient. Thus

$$(1+x)^7 = 1 + 7x + 21x^2 + 35x^3 + 35x^4 + 21x^5 + 7x^6 + x^7$$

1.3

The second term of the binomial x can be made negative or replaced by more difficult terms.

$$\begin{aligned} (1-x)^7 &= 1 + 7(-x) + 21(-x)^2 + 35(-x)^3 + 35(-x)^4 + 21(-x)^5 + \\ & \quad + 7(-x)^6 + (-x)^7 \\ &= 1 - 7x + 21x^2 - 35x^3 + 35x^4 - 21x^5 + 7x^6 - x^7 \end{aligned}$$

When x is replaced by a term such as $2y$ care must be taken to see that each part is raised to the power of x, for example x^3 becomes $(2y)^3$ which is $8y^3$. The bracket is sometimes forgotten, giving the incorrect answer $2y^3$. Bearing this in mind:

$$(1+2y)^4 = 1 + \frac{4}{1}.(2y) + \frac{4.3}{1.2}.(2y)^2 + \frac{4.3.2}{1.2.3}.(2y)^3 + \frac{4.3.2.1}{1.2.3.4}.(2y)^4$$

$$= 1 + 4.2y + 6.4y^2 + 4.8y^3 + 1.16y^4$$

$$= 1 + 8y + 24y^2 + 32y^3 + 16y^4$$

Exercise 1b. Expand by the binomial theorem:

1. $(1-x)^9$ 2. $(1+2y)^6$ 3. $(1-x)^5$
4. $(1+3y)^4$ 5. $(1-2x)^6$ 6. $(1-3x)^4$

1.4

If x is made much smaller than one its higher powers will become very small and so the binomial theorem can be used for finding quite close approximations.

Suppose we wish to find $(0·99)^8$ correct to five significant figures. As a binomial this must be $(1-0·01)^8$. We therefore need the expansion of $(1-x)^8$ which is $1-8x+28x^2-56x^3+70x^4-56x^5+28x^6-8x^7+x^8$. We now substitute $x = 0·01$ going as far as necessary, i.e. six decimal places (for the answer will be about one).

1	1·000000	
$-8(0·01)$		$-0·080000$
$+28(0·0001)$	0·002800	
$-56(0·000001)$		$-0·000056$
$+70(0·00000001)$	0·00000070	
adding columns	1·002800	$-0·080056$

ignoring the fifth term which is too small. There is no need to go further for the final four terms will be even smaller than the fifth. Subtracting gives 0·922744. To five significant figures this gives $(0·99)^8 = 0·92274$.

Exercise 1c

1. Expand $(1-x)^6$ and use the expansion to evaluate $(0·98)^6$ correct to three significant figures.
2. Using the expansion of $(1+x)^5$ find the value of $(1·01)^5$ correct to five significant figures.
3. Using a binomial expansion find the value of $(1·01)^8$ correct to five decimal places.
4. Expand $(1-2x)^6$ and then use the expansion to evaluate $(0·98)^6$ correct to three significant figures. (Compare with Question 1.)

1.5

It is now time to replace the 1 in the binomial by a general letter a. The expansion then becomes:

$$(a+x)^n = a^n + \frac{n}{1}.a^{(n-1)}.x + \frac{n.(n-1)}{1.2}.a^{(n-2)}.x^2 +$$

$$+ \frac{n(n-1)(n-2)}{1.2.3}.a^{(n-3)}.x^3 +$$

$$+ \frac{n(n-1)(n-2)(n-3)}{1.2.3.4}.a^{(n-4)}.x^4 \ldots x^n$$

It is a help to notice that the power of a and the power of x of each term add up to n. Using the result to expand $(a+x)^6$ we obtain

$$a^6+6a^5.x+15a^4.x^2+20a^3.x^3+15a^2.x^4+6a.x^5+x^6.$$

It can also be applied to harder examples, i.e.

$$(a-2y)^6 = a^6+6.a^5(-2y)+\frac{6.\overset{3}{5}}{1.2}.a^4(-2y)^2+\frac{6.5.4}{1.2.3}.a^3(-2y)^3+$$

$$+\frac{6.5.4.\overset{3}{3}}{1.2.3.\overset{}{4}}.a^2(-2y)^4+\frac{6.5.4.3.2}{1.2.3.4.5}.a(-2y)^5+$$

$$+\frac{6.5.4.3.2.1}{1.2.3.4.5.6}.(-2y)^6$$

$$= a^6-12a^5y+60a^4y^2-160a^3y^3+240a^2y^4-192ay^5+64y^6$$

Exercise 1d. Expand by the binomial theorem:

1. $(a+x)^5$ 2. $(a+x)^8$ 3. $(a-x)^5$

4. $(a+3y)^4$ 5. $(2a+x)^6$ 6. $(2b-3y)^4$

1.6

If a particular term is required it is easiest, at this stage, to write down the series without any simplification until the required term is reached.

For example let us find the term independent of x in the expansion of $\left(x-\frac{2}{x}\right)^6$.

$$\left(x-\frac{2}{x}\right)^6 = x^6+6.x^5\left(-\frac{2}{x}\right)+\frac{6.\overset{3}{5}}{1.2}.x^4\left(-\frac{2}{x}\right)^2+\frac{6.5.4}{1.2.3}.x^3\left(-\frac{2}{x}\right)^3\ldots$$

The fourth term is the required one, i.e.

$$20x^3.\frac{-8}{x^3}$$

This gives the answer as -160.

Exercise 1e

1. Find the third term in the expansion $(a-2b)^{10}$.

2. Find the term containing a^6 in the expansion of $(a^2-2x)^7$.

3. Find the term independent of y in the expansion of $\left(y^2-\frac{2}{y}\right)^6$.

4. Find the term in y in the expansion of $\left(y-\frac{2}{y}\right)^5$.

5. Find the middle term in the expansion of $\left(x - \dfrac{1}{y}\right)^8$.

6. Find the coefficient of x^2 in the expansion of $\left(2 + \dfrac{1}{x}\right)^2 \cdot \left(1 - \dfrac{x}{2}\right)^7$.

2. Differentiation of Powers of x

2.1

The first aim in Calculus is to find the rate of growth of a function of a variable, which sounds much more complicated than it really is. Let us take an equation such as $y = 2x^2 - 5x - 7$. This is what we call a function of x, which is the variable. As x changes the value of y changes and a graph gives the picture of the change. Instead of y it is sometimes written as function of x, i.e. $f(x)$ or $F(x)$. A function of x is therefore any expression containing powers of x and added or multiplied constants.

A very simple function of x is the area of a square side x. The area $(y) = x^2$. Our aim is to find the rate of growth of y if x changes by an infinitesimal amount. Let us start by allowing x to grow slightly by a small amount which is called δx, then y will also change by a small amount which can be called δy. The new area is $y + \delta y$ and since it was caused by the increase in x it must equal $(x + \delta x)^2$.

$$\therefore y + \delta y = (x + \delta x)^2 = x^2 + 2x \cdot \delta x + (\delta x)^2$$

and is represented by the largest square in the diagram, although δx has had to be made rather large in order that it can be seen clearly. A small increase in x should be thought of as $\frac{1}{1000}$ of x $(0 \cdot 1\%)$ or smaller. It is called a small increment. From the diagram it is seen that the whole figure is made up of a larger square of area x^2, a very very small square $(\delta x)^2$ and the two shaded rectangles each of area $x \cdot \delta x$. The whole figure is $y + \delta y$.

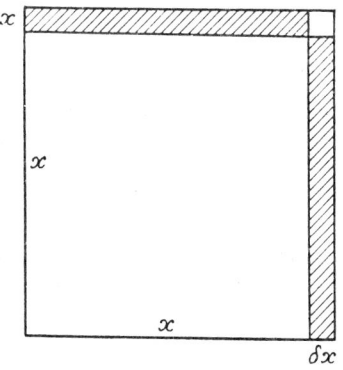

$$y + \delta y = x^2 + 2x \cdot \delta x + (\delta x)^2$$

For the original area $y = x^2$ and this can be subtracted from the equation giving $\delta y = 2x \cdot \delta x + (\delta x)^2$.

The rate of growth of δy with δx is therefore:

$$\frac{\delta y}{\delta x} = 2x + \delta x$$

Now the smaller we make δx the nearer we get to our aim of finding the rate of growth at the exact area $y = x^2$. From the right-hand side of the equation $\frac{\delta y}{\delta x} = 2x + \delta x$ the smaller we make δx the nearer we come to $2x$. Let us here make a reasonable assumption that when we reach the limit of making δx infinitesimal we come to the exact rate of growth which we shall call $\frac{dy}{dx}$. Thus when $y = x^2$ the rate of growth $\frac{dy}{dx} = 2x$.

Let us now apply the same argument to a cube of volume $y = x^3$. If x increases a small amount δx, causing a volume change δy then

$$y + \delta y = (x + \delta x)^3 = x^3 + 3x^2 \cdot \delta x + 3x \cdot (\delta x)^2 + (\delta x)^3$$

Subtracting the original cube $y = x^3$ we obtain:

$$\delta y = 3x^2 \cdot \delta x + 3x(\delta x)^2 + (\delta x)^3$$

i.e.
$$\frac{\delta y}{\delta x} = 3x^2 + 3x \cdot \delta x + (\delta x)^2$$

The smaller we make δx the nearer both $3x \cdot \delta x$ and $(\delta x)^3$ come to equalling zero. We can again apply our reasonable assumption and reach the limit of making δx infinitesimal to give $\frac{dy}{dx} = 3x^2$. The rate of growth $\frac{dy}{dx}$ is called the first derivative or the differential coefficient of y with respect to x, and is the ratio of two infinitesimally small terms. Other forms of the first derivative are $f'(x)$, $F'(x)$ or, by replacing y by the original equation, $\frac{d(x^2)}{dx}$, $\frac{d(x^3)}{dx}$, etc.

Exercise 2a. Find, using the results of Chapter 1 to expand the powers of $(x + \delta x)$, the differential coefficients of the following functions:

1. $y = x^4$
2. $y = x$
3. $y = x^5$
4. $y = x^8$

2.2

Fractional and negative powers of x can be dealt with by the same method.

Let us take $y = \sqrt{x}$, i.e. $y = x^{1/2}$

then $$y + \delta y = (x + \delta x)^{1/2}$$

By the binomial theorem –

$$y + \delta y = x^{1/2} + \tfrac{1}{2} . x^{-1/2} . dx + \frac{\tfrac{1}{2} . -\tfrac{1}{2}}{1 . 2} . x^{-3/2} . (\delta x)^2$$

$$+ \text{terms in higher powers of } \delta x$$

Subtracting the original equation:

$$\delta y = \tfrac{1}{2} . x^{-1/2} . \delta x - \tfrac{1}{8} . x^{-3/2} . (\delta x)^2 + \text{terms in higher powers of } \delta x$$

i.e. $\dfrac{\delta y}{\delta x} = \tfrac{1}{2} . x^{-1/2} - \tfrac{1}{8} . x^{-3/2} . \delta x + \text{terms in higher powers of } \delta x$

As δx becomes infinitesimal the second and all further terms tend to zero and the limit gives:

$$\frac{dy}{dx} = \tfrac{1}{2} . x^{-1/2} \quad \text{or} \quad \frac{1}{2\sqrt{x}}$$

Repeating the process for $y = \dfrac{1}{x}$, i.e. $y = x^{-1}$

$$y + \delta y = (x + \delta x)^{-1}$$

$$= x^{-1} + (-1) . x^{-2} . \delta x + \frac{-1 . -2}{1 . 2} . x^{-3} . (\delta x)^2$$

$$+ \text{terms in higher powers of } \delta x$$

Subtracting the original equation:

$$\delta y = -x^{-2} . \delta x + x^{-3}(\delta x)^2 + \text{terms in higher powers of } \delta x$$

i.e. $\dfrac{\delta y}{\delta x} = -x^{-2} + x^{-3} . \delta x + \text{terms in higher powers of } \delta x$

Applying the limit:

$$\frac{dy}{dx} = -x^{-2} \quad \text{or} \quad -\frac{1}{x^2}$$

Exercise 2b. Find the differential coefficients of the following functions:

1. $y = x^{1/3}$ 2. $y = x^{-2}$ 3. $y = x^{1/4}$

4. $y = x^{-3}$ 5. $y = x^{-1/2}$ 6. $y = x^{-1/3}$

B

2.3

The general formula to cover every case so far considered is $y = x^n$.
Applying the same process as before $y + \delta y = (x + \delta x)^n$, i.e.

$$y + \delta y = x^n + n.x^{n-1}.\delta x + \frac{n.(n-1)}{1.2}.x^{n-2}.(\delta x)^2$$

$$+ \text{terms in higher powers of } \delta x$$

Subtracting the original equation:

$$\delta y = n.x^{n-1}.\delta x + \frac{n.(n-1)}{1.2}.x^{n-2}(\delta x)^2 + \text{higher terms in } \delta x$$

i.e. $\dfrac{\delta y}{\delta x} = n.x^{n-1} + \frac{1}{2}.n.(n-1)x^{n-2}.\delta x + \text{higher terms in } \delta x$

Applying the limit $\dfrac{dy}{dx} = n.x^{n-1}$ and it will be seen that every answer
we have obtained fits this general formula.

When $\qquad\qquad y = x^n \qquad \dfrac{dy}{dx} = n.x^{n-1}$

This applies for any value of n except $n = 0$, when the equation reduces
to $y = 1$ (constant).

Exercise 2c. Applying the general formula find the derivatives of:

1. x^9	2. x^{27}	3. x^{-7}	4. x^{-9}
5. $x^{1/5}$	6. $x^{1/6}$	7. $x^{2/5}$	8. $x^{3/7}$
9. $x^{-1/5}$	10. $x^{-2/5}$	11. $x^{-3/7}$	12. $x^{0.6}$

3. The Trigonometrical Ratios

3.1

Since the circumference of a circle of radius r is given by $2\pi r$ it is
useful to call the angle at the centre 2π. This new measure of angle is
in radians, and so we have $360° = 2\pi$ radians. To find the length of
any arc of a circle it is only necessary to multiply the radius by the
angle subtended at the centre measured in radians.

For conversions we use the fact that two right angles will be $180°$, or
π radians.

To change from degrees to radians we therefore have to multiply by
$\dfrac{\pi}{180}$. This gives $270°$ as $\overset{3}{\underset{\cancel{270}}{270}} \times \dfrac{\pi}{\cancel{180}} = \dfrac{3\pi}{2}$ radians.

To change from radians to degrees the multiplying factor will be $\dfrac{180}{\pi}$. This gives $\dfrac{\pi}{5}$ radians as $\dfrac{\overset{36}{\cancel{\pi}}}{5} \times \dfrac{\overset{36}{180}}{\cancel{\pi}} = 36°$.

Since 2π represents an angle of four right angles it can be added to or subtracted from, an angle without changing the position of the angle in the same way that 360° can be added or subtracted. Radians are said to be circular measure.

Exercise 3a. Convert into radians:

1. 90°	2. 45°	3. 60°	4. 135°
5. 300°	6. 120°	7. 405°	8. 10°

Convert into degrees:

9. $\dfrac{\pi}{4}$	10. $\dfrac{\pi}{6}$	11. $\dfrac{2\pi}{3}$	12. 3π
13. $\dfrac{\pi}{10}$	14. $\dfrac{2\pi}{5}$	15. $\dfrac{4\pi}{9}$	16. $\dfrac{31\pi}{18}$

3.2

In a right-angled triangle the hypotenuse is always fixed as the longest side, i.e. that opposite the right angle. The angle α is made up of the hypotenuse on one side, and the adjacent side on the other. The third side which does not make part of the angle α is called the opposite side. The three basic ratios are:

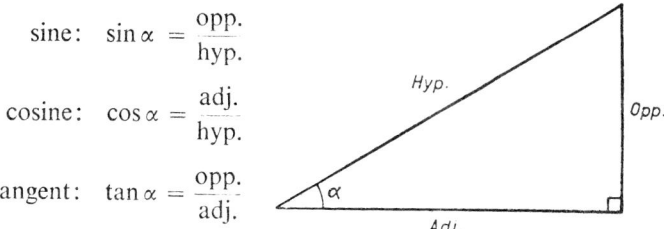

$$\text{sine:} \quad \sin \alpha = \frac{\text{opp.}}{\text{hyp.}}$$

$$\text{cosine:} \quad \cos \alpha = \frac{\text{adj.}}{\text{hyp.}}$$

$$\text{tangent:} \quad \tan \alpha = \frac{\text{opp.}}{\text{adj.}}$$

We now define the remaining three pairs:

$$\text{secant:} \quad \sec \alpha = \frac{\text{hyp.}}{\text{adj.}} = \frac{1}{\cos \alpha}$$

$$\text{cosecant:} \quad \operatorname{cosec} \alpha = \frac{\text{hyp.}}{\text{opp.}} = \frac{1}{\sin \alpha}$$

$$\text{cotangent:} \quad \cot \alpha = \frac{\text{adj.}}{\text{opp.}} = \frac{1}{\tan \alpha}$$

3.3

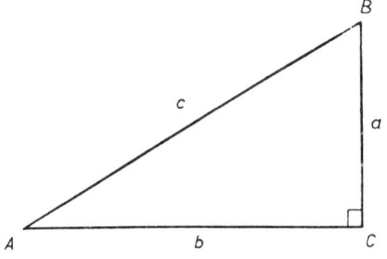

We shall now consider a right-angled triangle labelled in the standard way, with small letters representing the sides opposite the angle marked with the same capital letter.

Since this is a right-angled triangle, and the angles of any triangle add up to 180°, we see that $A° + B° = 90°$.

By definition $\sin A = \dfrac{a}{c}$ and $\cos B = \dfrac{a}{c}$, i.e. $\sin A = \cos B$; or $\sin A° = \cos(90 - A)°$. Thus $\sin 60° = \cos 30°$.

E.g. $\sin x° = \cos 2x°$, where $x°$ is acute, but $\sin x° = \cos(90 - x)°$, i.e. $90° - x° = 2x°$, which gives $x° = 30°$.

By definition $\tan A = \dfrac{a}{b}$ and $\cot B = \dfrac{a}{b}$, thus $\tan A = \cot B$, i.e. $\tan A° = \cot(90 - A)°$. E.g. $\tan 60° = \cot 30°$.

Exercise 3b

1. Show that $\cos A° = \sin(90 - A)°$ and change $\cos 30°$, $\cos 75°$, and $\cos 45°$ into sines.

2. Show that $\cot A° = \tan(90 - A)°$ and change $\cot 15°$, $\cot 60°$ and $\cot 90°$ into tangents.

3. Show that $\sec A° = \operatorname{cosec}(90 - A)°$ and change $\sec 20°$, $\sec 55°$ and $\sec 80°$ into cosecants.

4. Show that $\operatorname{cosec} A° = \sec(90 - A)°$ and change $\operatorname{cosec} 21°$, $\operatorname{cosec} 45°$ and $\operatorname{cosec} 76°$ into secants.

3.4

Using the same right-angled triangle we can build up some simple relationships for the different ratios of an angle.

(a)
$$\sin A = \frac{a}{c} \quad \text{and} \quad \cos A = \frac{b}{c}$$

$$\therefore \frac{\sin A}{\cos A} = \frac{\dfrac{a}{c}}{\dfrac{b}{c}} = \frac{a}{b} = \tan A$$

i.e. $\tan A = \dfrac{\sin A}{\cos A}$.

(b) If we wish to square $\sin A$, i.e. $(\sin A)^2$ it is written as $\sin^2 A$.

$$\therefore \sin^2 A + \cos^2 A = \frac{a^2}{c^2} + \frac{b^2}{c^2} = \frac{a^2 + b^2}{c^2} = \frac{c^2}{c^2} = 1$$

(because by Pythagoras' theorem $a^2 + b^2 = c^2$), i.e. $\sin^2 A + \cos^2 A = 1$.

(c) $$\sec^2 A - \tan^2 A = \frac{c^2}{b^2} - \frac{a^2}{b^2} = \frac{c^2 - a^2}{b^2} = \frac{b^2}{b^2} = 1$$

i.e. $1 + \tan^2 A = \sec^2 A$.

Exercise 3c. By similar reasoning show:

1. $\cot A = \dfrac{\cos A}{\sin A}$.

2. $\sin^2 B + \cos^2 B = 1$.

3. $1 + \cot^2 A = \operatorname{cosec}^2 A$.

4. $\sec^2 A + \operatorname{cosec}^2 A = \sec^2 A \cdot \operatorname{cosec}^2 A$ (start with left-hand side and put it over L.C.M. $a^2 \cdot b^2$).

3.5

The ratios of $45°$, $30°$ and $60°$ are used so often, especially in mechanics, that they are treated as standard in the surd form (i.e. with the use of a square root sign).

To build up the $45°$ ratios we think of an isosceles right-angled triangle, with the two equal sides of length unity. Then by Pythagoras' theorem the square of the hypotenuse equals $1^2 + 1^2 = 2$. The length of the hypotenuse is therefore $\sqrt{2}$.

$$\sin 45° = \cos 45° = \frac{1}{\sqrt{2}} \quad \text{and} \quad \tan 45° = 1$$

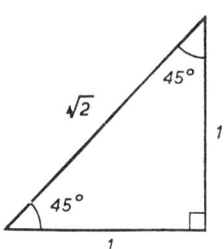 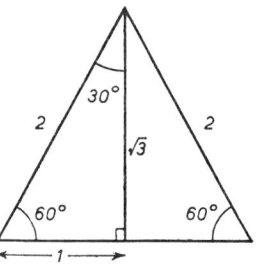

To build up the $30°$ and $60°$ ratios we consider an equilateral triangle of side two units. The perpendicular on to the one side bisects that side and the angle, giving a smaller triangle of $30°$, $60°$, $90°$. By Pythagoras' theorem the square of the altitude is $2^2 - 1^2 = 4 - 1 = 3$. The length of the altitude is $\sqrt{3}$.

$$\sin 30° = \cos 60° = \tfrac{1}{2}$$

$$\cos 30° = \sin 60° = \frac{\sqrt{3}}{2}$$

$$\tan 30° = \frac{1}{\sqrt{3}} \quad \text{and} \quad \tan 60° = \sqrt{3}$$

Exercise 3d. From the diagrams find the values of:

1. $\sec 45°$	2. $\cot 45°$	3. $\cot 30°$
4. $\sec 30°$	5. $\operatorname{cosec} 30°$	6. $\operatorname{cosec} 60°$
7. $\cot 60°$	8. $\operatorname{cosec} 45°$	9. $\sec 60°$

4. Graphical Representation of a Curve; Gradients

4.1

In co-ordinate or Cartesian geometry we take two axes perpendicular to each other. The horizontal axis is the x-axis which is positive to the

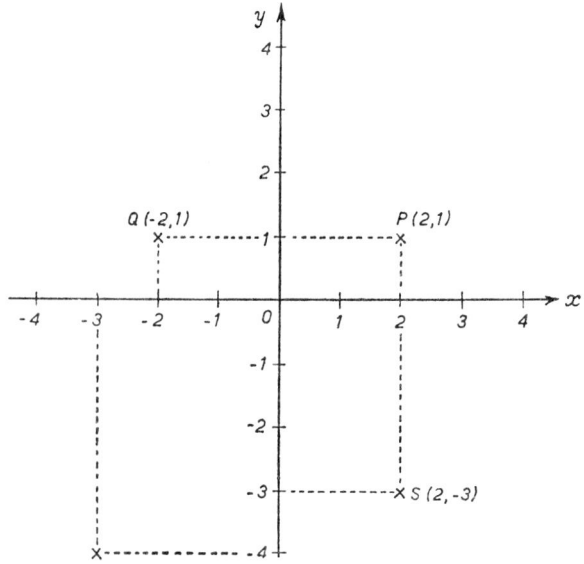

right. The vertical axis is the y-axis which is positive upwards. The two axes cross at the origin of co-ordinates O. Any point P can be fixed in two dimensions by stating its x co-ordinate or abscissa and then its y co-ordinate or ordinate. These are written in a bracket (x, y).

In the diagram the point P is in the first quadrant and is $(2,1)$. Q is in the second quadrant and is $(-2,1)$. In the third quadrant both co-ordinates are negative and $(-3, -4)$ brings us to point R. In the fourth quadrant the x co-ordinate has become positive to give point S as $(2, -3)$.

The x-axis is the line $y = 0$, and the y-axis the line $x = 0$. The origin is $(0,0)$.

Exercise 4a. Plot the following Cartesian co-ordinates:

1. $(3,5)$	2. $(-3,5)$	3. $(3,-5)$	4. $(-3,-5)$
5. $(2,1)$	6. $(-2,-3)$	7. $(-1,1)$	8. $(2,-2)$
9. $(3,3)$	10. $(4,-1)$	11. $(-1,4)$	12. $(-4,1)$

4.2

If we now take the simple function $y = x^2$ of section 2.1 we can plot it by obtaining the ordinates for certain values of x.

x	-3	-2	-1	$-\frac{1}{2}$	0	$\frac{1}{2}$	1	2	3
y	9	4	1	$\frac{1}{4}$	0	$\frac{1}{4}$	1	4	9

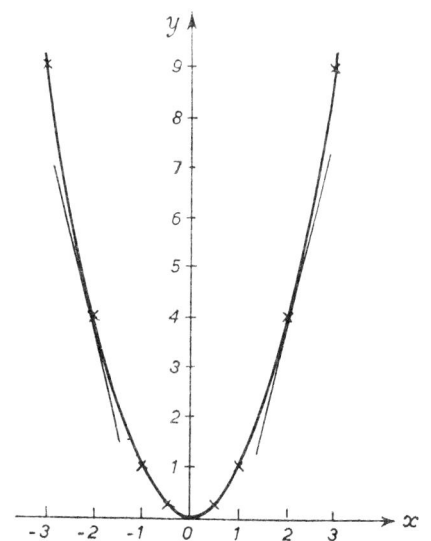

This curve is a parabola and is the shape obtained from any quadratic equation.

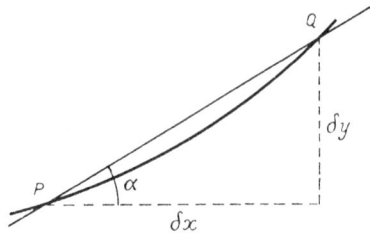

Let us now consider a much enlarged portion of the curve in the region $x = \frac{1}{2}$. We draw a small chord PQ. The change in the x co-ordinate from P to Q will be very small and can be represented by δx. The change in the y co-ordinate will be δy. The slope or gradient of the chord $\tan \alpha$ is given by $\dfrac{\delta y}{\delta x}$. If Q is now brought nearer to P the chord becomes shorter and δx smaller. This process can be taken to its limit when Q coincides with P, which makes the line PQ into the tangent. At the same time δx has reached its limit of becoming infinitesimal and so the rate of change has become $\dfrac{dy}{dx}$, the differential coefficient.

Thus the rate of growth of a function of x, at the point x, is equal to the slope or gradient of the tangent at that point.

For example, find the slope of the tangent of the curve $y = x^2$ at the point (a) $x = 2$, (b) $y = 0$, (c) $x = -2$.

$y = x^2$ and applying the result of section 2.1 the slope $\dfrac{dy}{dx} = 2x$.

(a) When $x = 2$, slope $= +4$.

(b) When $y = 0$, then $x = 0$, for they are connected by $y = x^2$.
$\therefore$ slope $= 0$.

(c) When $x = -2$, slope $= -4$.

The $+$ve sign means the tangent is sloping forwards, the zero slope means it is horizontal and the $-$ve sign is for a backwards slope as shown on the graph of $y = x^2$. $\tan \alpha$ is often put equal to m and so the slope is called the m of the line.

Exercise 4b

1. Plot the curve $y = x^3$ between $x = -2$ and $x = +2$. Find the gradients of the tangents at the points $x = -2$ and $y = +8$.

2. Plot the curve $y = \dfrac{1}{x}$ between $x = -6$ and $x = +6$ (except between $x = -\frac{1}{10}$ and $x = +\frac{1}{10}$). Find the slopes of the tangents at $x = -2$ and $x = +2$.

3. Plot the curve $y = \sqrt{x}$ between $x = -1$ and $+9$ and find the gradients at the points where $y = 2$ and $y = -2$.

4. Plot the curve $y = \dfrac{1}{x^2}$ between $x = -6$ and $x = 6$ (except between $x = -\frac{1}{4}$ and $x = +\frac{1}{4}$).

4.3

A special line is the straight line for it has a constant slope. Let us name this slope m and make the intercept distance on the y-axis equal to c as shown.

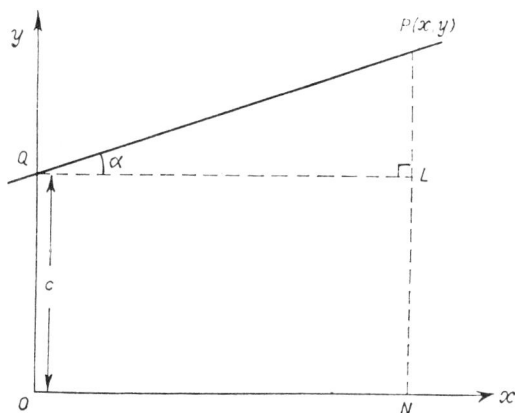

Taking any point $P(x, y)$ on the straight line if we can connect x and y we will obtain the equation of the line. $PN = y$ and $LN = c$, since $OQLN$ is a rectangle. $\therefore PL = PN - LN = y - c$. $QL = x$.

$$\text{Tan } \alpha = m = \frac{PL}{QL} = \frac{y - c}{x},$$

$\therefore mx = y - c$. This gives the equation to a straight line as

$$y = mx + c,$$

where m is the slope and c is the intercept distance on the y-axis.

Thus we see that an equation of the first degree (containing only x's y's and constants without any powers of x or y, or any xy terms) represents a straight line and so it is called a linear equation.

Thus if we are given a linear equation $5y = 4x - 10$ we can find the slope and intercept distance by making y the subject, i.e.

$$y = \tfrac{4}{5}.x - 2.$$

Comparing $y = mx + c$ we obtain $m = \tfrac{4}{5}$ and $c = -2$.

Exercise 4c. Obtain the gradient and intercept distances of the following straight lines:

1. $2y = 3x + 4$ 2. $6y + 4 = 3x$ 3. $3y + 2x - 3 = 0$

4. $3x = 2y + 1$ 5. $3x + 5 = 2y$ 6. $x = \dfrac{y+3}{2}$

4.4

If a point lies on a curve its co-ordinates must satisfy the equation of the curve. Let us consider if the point $(1, 2)$ lies on the curve

$$y = 3x^2 + 2x - 6.$$

Let us put the value $x = 1$ in the equation and obtain

$$y = 3 + 2 - 6 = -1.$$

Our value for $x = 1$ is $y = 2$, so it does not satisfy the equation (i.e. it does not work out to $y = 2$, but to $y = -1$), and so $(1, 2)$ is not on the line.

If we are told the points $(1, 2)$ and $(3, -3)$ lie on a straight line we can find the equation of that line. Let the equation be $y = mx + c$.

From the first point:

$$2 = m \times 1 + c \quad \text{i.e. } 2 = m + c \tag{1}$$

From the second point:

$$-3 = m \times 3 + c \quad \text{i.e. } -3 = 3m + c \tag{2}$$

Subtracting (1) from (2), $-5 = 2m$, $\therefore m = \dfrac{-5}{2}$

Then substituting this value in (1),

$$c = 2 + \frac{5}{2} = \frac{9}{2}$$

Therefore the equation to the line is

$$y = \frac{-5x}{2} + \frac{9}{2}$$

i.e. $2y = -5x + 9$, or $5x + 2y - 9 = 0$.

As a check we see that both points satisfy this equation.

Exercise 4d

1. Find the value of y if point $(2, y)$ lies on $y = x^3 + 2x^2 + 3$.
2. Find the values of x if points $(x, 1)$ lie on $y = 2x^2 + 5x + 4$.
3. Which of the points $(-1, 8)$, $(0, 3)$, $(2, 1)$ and $(1, 0)$ lie on the curve $y = 3x^2 - 4x + 1$.
4. Find the slope of the straight line which passes through $(0, 0)$ and $(4, 4)$.
5. Find the equation of the straight line which passes through $(2, -1)$ and $(3, 2)$.
6. A straight line has slope $m = \frac{1}{2}$ and passes through $(6, -1)$. Find its equation. Does it pass through the origin?

5. Differentiation of Sums and Products

5.1

Let us consider the equation $y = x(1 + x)$, multiplying out $y = x + x^2$. If x increases by δx then

$y + \delta y = (x + \delta x) + (x + \delta x)^2 \quad \therefore y + \delta y = x + \delta x + x^2 + 2x \cdot \delta x + (\delta x)^2$

Subtracting the original: $\delta y = \delta x + 2x \cdot \delta x + (\delta x)^2$

i.e.
$$\frac{\delta y}{\delta x} = 1 + 2x + \delta x$$

If we make δx tend to an infinitely small value we obtain $\dfrac{dy}{dx} = 1 + 2x$.

This demonstrates the general rule that sums or differences of terms can be differentiated as if they are separate terms, for the derivative of x is 1 and of x^2 is $2x$.

5.2

A multiplied constant will carry through the differentiation. Let $y = ax^2$ then

$$y + \delta y = a(x + \delta x)^2$$
$$y + \delta y = ax^2 + 2ax \cdot \delta x + a(\delta x)^2$$
$$\delta y = 2ax \cdot \delta x + a(\delta x)^2$$
$$\frac{\delta y}{\delta x} = 2ax + a\delta x$$

In the limit $\dfrac{dy}{dx} = 2ax$, or $a \cdot 2x$.

5.3

An added constant will vanish after differentiation.

Let $\qquad\qquad\qquad y = x^2 + c$

then $\qquad\qquad y + \delta y = (x + \delta x)^2 + c$

$\qquad\qquad\qquad y + \delta y = x^2 + 2x \cdot \delta x + (\delta x)^2 + c$

Subtracting the original:

$$\delta y = 2x \cdot \delta x + (\delta x)^2$$

$$\therefore \frac{\delta y}{\delta x} = 2x + \delta x$$

In the limit $\dfrac{dy}{dx} = 2x$.

If we apply these two results to the straight line $y = mx + c$ we obtain

$$\frac{dy}{dx} = m \cdot 1 + \text{zero}$$

i.e. $\dfrac{dy}{dx} = m$, the constant slope, which confirms the result of section 4.3.

We can now differentiate the first equation of section 2.1.

$$y = 2x^2 - 5x - 7$$

$$\frac{dy}{dx} = 2 \cdot 2x - 5$$

$$\frac{dy}{dx} = 4x - 5$$

From this we find the slope of the tangent when $x = 0$ is

$$m = 4 \times 0 - 5 = -5$$

by the result of section 4.2. When $x = 2$ the m is $8 - 5 = 3$.

If we wish to find the gradient where the curve cuts the x-axis we have to solve the curve's equation with that of the x-axis, i.e. $y = 0$

$$y = 2x^2 - 5x - 7 \quad \text{and} \quad y = 0$$

$$\therefore 0 = 2x^2 - 5x - 7$$

$$0 = (2x - 7)(x + 1)$$

$$x = +\frac{7}{2} \quad \text{or} -1$$

When $x = +\dfrac{7}{2}$, slope $m = \dfrac{4 \cdot 7}{2} - 5 = 9$.

When $x = -1$, slope $m = 4 \cdot (-1) - 5 = -9$.

Exercise 5a. Differentiate with respect to x:

1. $x^4 + x^3$

2. $x - \sqrt{x}$

3. $x^4 + \dfrac{1}{x^4}$

4. $6x^2 - 7$

5. $7x^3 + 7 + \dfrac{2}{x}$

6. $\dfrac{x^3}{3} - \dfrac{1}{2x^2}$

7. $6x^2 + 5x + 2$

8. $x^2 + 4 \cdot x^{-3}$

9. $7x^{3/7}$

10. Find the slopes at the two points where $y = x^2 - 3x - 4$ cuts the x-axis.

11. Find the slopes at the three points where $y = x^3 - x^2 - 2x$ cuts the x-axis (note the factor x).

12. Find the gradients at the points where $y = 4 - x^2$ cuts the x-axis.

5.4

Let us consider the product $y = u \cdot v$, where u and v are functions of x. Then if x increases by δx, u increases by δu, v by δv and y by δy.

$$\therefore \ y + \delta y = (u + \delta u)(v + \delta v)$$

$$\cancel{y} + \delta y = u\cancel{v} + u \cdot \delta v + v \cdot \delta u + \delta u \cdot \delta v$$

$$\frac{\delta y}{\delta x} = u \cdot \frac{\delta v}{\delta x} + v \cdot \frac{\delta u}{\delta x} + \delta u \cdot \frac{\delta v}{\delta x}$$

In the limit when δx becomes infinitesimal so does δu, therefore $\delta u \cdot \dfrac{\delta v}{\delta x}$ becomes zero; and

$$\frac{dy}{dx} = u \cdot \frac{dv}{dx} + v \cdot \frac{du}{dx}$$

We know that if $y = x^6$, $\dfrac{dy}{dx} = 6x^5$. We shall use this to check the product formula:

$$y = x^2 \cdot x^4$$

then
$$u = x^2; \quad v = x^4$$

$$\frac{du}{dx} = 2x; \quad \frac{dv}{dx} = 4x^3$$

$$\therefore \ \frac{dy}{dx} = x^2 \cdot 4x^3 + x^4 \cdot 2x$$

$$= 4x^5 + 2x^5 = 6x^5$$

After some practice this can be done without putting in the middle steps with u and v. The derivative of a product is the first part multiplied by the derivate of the second, plus the second multiplied by the derivative of the first. Thus if

$$y = (3x^2 + 2)(2x^4 + 3x)$$

$$\frac{dy}{dx} = (3x^2 + 2)(8x^3 + 3) + (2x^4 + 3x)(6x)$$

$$= 24x^5 + 16x^3 + 9x^2 + 6 + 12x^5 + 18x^2$$

$$= 36x^5 + 16x^3 + 27x^2 + 6$$

If we had multiplied out the product we obtain

$$y = 6x^6 + 4x^4 + 9x^3 + 6x$$

then $$\frac{dy}{dx} = 36x^5 + 16x^3 + 27x^2 + 6$$

We shall practise the product rule here to get used to it, as we shall not always be able to multiply out.

Exercise 5b. Differentiate with respect to x (as products):

1. x^6 as (i) $x^3 . x^3$; (ii) $x . x^5$
2. $(6x + 5)(3x + 2)$
3. $(4x^2 + 1)(x + 1)$
4. $x(6x^2 + 7x + 2)$
5. $(6x^2 + 7x)(2 - x)$
6. $\left(1 + \dfrac{1}{x}\right)(2 + x)$
7. $(3x^2) . \sqrt{x}$
8. Find the gradients at the three points where $y = (x^2 - 4)(x + 3)$ cuts the x-axis.

5.5

Successive differentiations with respect to x obey the rules already given.

$$y = 6x^3 + 7x + 3$$

then $$\frac{dy}{dx} = 18x^2 + 7$$

then $$\frac{d\left(\dfrac{dy}{dx}\right)}{dx} = 36x.$$

This is the second derivative and is written as $\dfrac{d^2 y}{dx^2}$ or $f''(x)$. If we require the third derivative, $\dfrac{d^3 y}{dx^3}$ or $f'''(x) = 36$. The process can be continued, unless the derivative vanishes as this one does for $f^{iv}(x) = 0$.

Exercise 5c. Find the second derivates for numbers 1 to 12 of Exercise 2c (continue from the answers giving the first derivatives).

13. Find the sixth derivative of $\dfrac{x^5}{4} + x^4 + 4$.

---

6. Choice and Chance

6.1

The first choice we shall consider is an arrangement which is a permutation. For an arrangement the order is important such as the gold, silver and bronze medal winners in the final of an Olympic race. With the other type of choice a selection is made. A group is chosen, but it does not matter about the order in which the choice is made, for example choosing six members for a committee. This is a combination.

Let us consider the six runners A, B, C, D, E, and F. Each man has a chance of winning, no matter how remote, so there are six possibilities for the winner. Which ever of these finishes first there will be five others still running. This means there are five possibilities of choosing the second. For any particular winner, say A, there are five ways of first and second finishing, A first then B, A then C, A then D, A then E, or A then F. For each of the other five possible winners there will be five possible sets of first and second. In all there are 30 different permutations of first and second, which is 6×5, the separate possibilities.

A first and second having passed the tape there are still four runners competing for third place. Thus for each of the possible 30 arrangements of the first two places there are four for third place. This means there are 120 permutations for the first three places. It is said that 120 is the permutation of three from six, or in our shorthand $^6P_3$.

If we now consider the possibilities for all six places we continue in the same way. Three runners can fill the fourth place, leaving two to fill the fifth place and then there is one runner who will be last. This gives a full arrangement $^6P_6 = 6.5.4.3.2.1 = 720$. Products of numbers, each 1 less than the previous, ending in 1 come into so many problems that a shorthand is used. $6.5.4.3.2.1$ is written 6! or $\lfloor 6$ (six factorial).

Exercise 6a

1. In a class of twenty pupils there is an election for form captain and deputy form captain. How many results are possible?
2. A cricket captain decides that his four bowlers should fill the last four places in the batting order. How many arrangements of the last four positions can he make?
3. How many four-letter codes can be made from the first ten letters of the alphabet if no letter is repeated?
4. Seven forwards, who can play in any position, are to fill the five forward positions of a hockey team. How many choices are there?
5. In how many ways can the letters of the word *bridge* be arranged?
6. How many arrangements can be made taking five of the letters of the word *education*?

6.2

We shall now consider permutations in which there is some condition that has to be satisfied. For example we may be asked to find the number of arrangements of the letters in the word *holiday* that begin with *h*. There is only one choice for the first letter as it must be *h*. There is a choice of six letters for second place, five for third place, etc., i.e. $1.6.5.4.3.2.1 = 720$.

Again we may have to find how many arrangements finish with a vowel. We can fill the places in any order, and so we start with the special condition, the final place. There are three vowels that can fill this position. To fill one of the other positions, we have a choice of six, the two vowels together with four consonants. The next place can be filled in five ways, etc. This gives $3.6.5.4.3.2.1 = 2160$ arrangements of *holiday* finishing in a vowel.

The problem may be worded so that we have to split it up into parts. We will find how many arrangements of the word *holiday* have their odd positions filled by consonants and even filled by vowels (*holiday* itself satisfies the condition). We start by filling the first, third, fifth, and seventh positions with the four consonants in $4.3.2.1 = 24$ ways. The second, fourth, and sixth positions can be filled by the three vowels in $3.2.1 = 6$ ways. Any set of vowels can be placed with any set of consonants to give the total arrangement as $6 \times 24 = 144$.

6.3

A special arrangement is the circular permutation. When five heads of state sit at a round table they choose it because the seats are equal in importance. The first man will sit down in one of the seats which he chooses at random. The positions now become important in relationship with the first man. The second man has four choices, the third man has three, etc. This gives a total of $1.4.3.2.1 = 24$ arrangements.

Exercise 6b

1. Find the number of arrangements of the letters of the word *object*. How many begin with *j*?
2. Arrange the word *reaction* with the consonants in the odd positions and vowels in the even positions. How many of these arrangements start with *t* and end with *a*?
3. How many three figure numbers can be made from the digits (*a*) 1, 2, 5, 7, (*b*) 0, 1, 2, 5, no digit being used more than once?
4. How many permutations are there of the letters of the word *thing* which do not end with the letters *ng* in that order?
5. A cricket team consists of five batsmen, two all-rounders and four bowlers. If the groups are to bat in this order, and each player can take any position within his own group, how many possible batting orders are there?
6. How many three-figure numbers can be formed from 0, 2, 5, 6, 8, 9, if none are repeated in any number? How many are odd?
7. Six children are going to play ring-a-roses. In how many ways can they join hands?
8. How many arrangements of the letters of the word *trend* will have *e* in the centre, if no letters are repeated?

6.4

A letter or digit may be repeated in the group to be arranged. If we have to make an arrangement of the word *paddle* we first of all assume that the two letters *d* are different, d_1 and d_2. This gives 6! arrangements.

Now let us consider two of the arrangements pad_1d_2le and pad_2d_1le. When we remove the distinguishing suffixes these become the same word. All the words would pair off to give these doubles, so we must divide by 2. This gives $6.5.4.3 = 360$. To find out how many words begin with *d* we have one choice for the first letter and that leaves us with five different letters *p, a, d, l, e* to fill five places. The arrangement is $1.5! = 120$.

If we had three letters alike d_1, d_2, d_3, there would have been six sets alike: d_1, d_2, d_3; d_1, d_3, d_2; d_2, d_1, d_3; d_2, d_3, d_1; d_3, d_1, d_2; and d_3, d_2, d_1 thus we must divide by 6, which is 3! In general if *r* letters are the same we must divide by *r*! For example if we arrange *success* we treat

c

them as all distinguishable to give 7! and then divide by 3!, because three letters are s, and 2! because two letters are c, i.e.

$$\frac{7.6.5.4.3.2.1}{3.2.1.2.1} = 420$$

6.5

Some conditions may mean that we have several completely separate problems. For example if we wish to find the number of three-letter words from the letters of *paddle* there are three problems:

(a) Without a d in the word we have a choice of three from *pale*, i.e. $4.3.2 = 24$.

(b) One letter d can fill any one of the three places, the other two places left being filled in 4.3 ways, i.e. $3 \times 4.3 = 36$.

(c) Two letters d can fill any two positions in $\frac{3.2}{2.1}$ ways. The third position can be filled in four ways giving $3 \times 4 = 12$.

The total arrangements are $24 + 36 + 12 = 72$.

6.6

For the final type of permutation positions may be repeated. If a bridge player holds an ace he may wish to find out how many arrangements there are for the other three aces dealt amongst the other three players. They may all be held by one player, they may have one each, or one player may have one ace and another two. In the last case it still has to be decided which is the single ace, and which player may have it. This small problem leads to so many sub-divisions that another approach is taken.

The ace of spades could have been dealt in three ways, to either of the three other players. The ace of clubs also could have been dealt in three ways, and the ace of hearts. The total arrangements are $3^3 = 27$.

In general if n objects can go to any of r people the number of arrangements is r^n.

Exercise 6c

1. In how many ways can five boys be arranged in order of merit? In how many arrangements will two particular boys be next to each other?

2. In how many ways can three prizes be given to six boys if a boy can win any number of prizes?

3. In how many ways can the letters of *rotor* be arranged? How many begin with t?

4. How many numbers greater than a thousand can be formed from the digits 0, 1, 2, 3, 4, and 5 if they can be used once only in any number?

5. How many four-letter words can be made from the letters of *erase*? How many begin with *e*? Repeat for three-letter words.

6. Find the number of arrangements of the words (*a*) LONDON, (*b*) DISS, (*c*) MATHEMATICAL.

7. Find the number of arrangements of the words (*a*) COMMITTEE, (*b*) BANANA.

8. In how many ways can six different books be divided amongst two boys? (This implies that each boy must always receive at least one book.)

6.7

For a combination we first of all make an arrangement and then divide by the number of groups which become the same when order no longer matters.

If we take the letters *A*, *B*, *C*, *D*, *E*, *F* the number of permutations of all six is 6!

But there is only one possible way to select all six, for we must take them all, that is we should divide the permutation by 6!

To arrange three letters from the six we have 6.5.4. Let us take any one arrangement, say *B*, *D*, *E*. The other arrangements that contain these letters are *B*, *E*, *D*; *E*, *D*, *B*; *E*, *B*, *D*; *D*, *B*, *E*; and *D*, *E*, *B*; making six in all. This is true of any group of three letters, so for a selection we must divide by 6, or 3!, which is again the factorial of the number of places to be filled. This is a general rule for combinations.

Thus the number of selections of three letters from the six is
$$\frac{6.5.4}{3.2.1} = 20.$$

Exercise 6d

1. How many triangles can be formed by joining any three of the vertices of a pentagon?

2. There are five boys and seven girls who are prefects. How many ways can they form a committee of four with the head girl (who is also one of the prefects) as chairwoman?

3. There are nine points in a plane, five of which are in a straight line. If no other sets of three or more points are in a line, how many triangles can be formed with the points as vertices?

4. Five boy prefects are available to be selected for three duties each morning. Six girl prefects are to share the three afternoon

duties. If a prefect does only one duty in any one day how many different rotas can be made?

5. A committee of three masters and two mistresses is to be formed from a school of seven masters and five mistresses. How many selections are there?

    If the twelve teachers include the Headmaster and Senior Mistress, how many committees will have one or other, but not both, of them included?

6. How many committees of three can be formed from six girls and six boys? In how many of these will the girls be in a majority?

6.8

To find the number of selections when any number of the objects can be taken we consider what happens to the objects. To find the number of selections of eight different novels we consider the first which may be taken or may be left, i.e. there are two selections. The same applies to the other seven giving a combined selection of 2^8 or 256. But this includes the case where each of the eight was not selected, so the total of selections of one or more book is 255.

6.9

The probability or chance of a thing happening is the ratio of the number of ways it can happen to the total number of choices.

In section 6.2 we found that the number of arrangements of *holiday* that begin with *h* comes to $1 \times 6!$ The total arrangements are $7!$, i.e. $7 \times 6!$ Thus the chance of beginning with *h* is 1 in 7.

Exercise 6e

1. How many selections can be made from ten different novels? (at least one must be selected).

2. What is the chance that an arrangement of the letters of *mathematical* begin with (i) *l*, (ii) *m*, and (iii) *a*?

3. What is the chance of a three-letter arrangement of *A*, *B*, *C*, *D*, *E*, *F* containing the letter *A*?

4. What is the chance of a three-letter selection of *A*, *B*, *C*, *D*, *E*, *F* containing the letter *A*?

5. How many selections may be made from five oranges? (at least one must be selected and they are of different quality).

6. How many selections may be made from three oranges and four apples taking at least one of each? (fruit differs in quality).

7. What is the chance of a four-letter arrangement of the letters of

river beginning with *r*? What is the chance for a three-letter arrangement?

8. In question 2 of Exercise 6d, what is the chance of the committee being all girls?

7. Angles of any Magnitude

7.1

The trigonometrical ratios of angles outside the range 0° to 90° or 0 to $\frac{\pi}{2}$ radians bear close relationships to ratios of angles within the range.

The best way of setting out any angle is to measure it from the horizontal right-hand line going round in an anti-clockwise direction for positive angles.

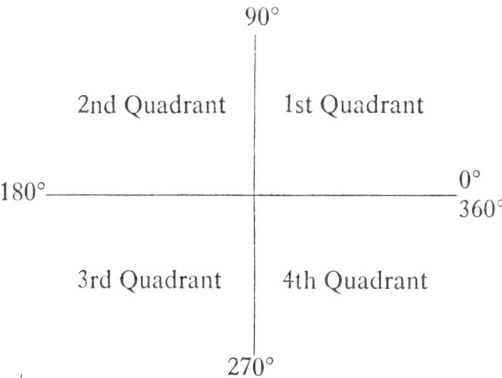

An angle between 0° and 90° will come in the first quadrant; 90° to 180° in the second; 180° to 270° in the third and 270° to 360° in the fourth.

Thus we can place any angle between 0° and 360° in one of the four quadrants. If an angle is outside this range we remember that turning through four right angles brings us back to the starting line, and the angles are equal. For example 10° is the same as 370°, or, subtracting four right angles, the same as − 350°. Any angle can be replaced by the equal angle between 0° and 360° by adding or subtracting any multiple of 360°. Thus 950° transforms to 950°–720°, i.e. 230°, and is in the third quadrant. In the same way $-120°$ transforms to $-120°+360°$, i.e. 240°, and is also in the third quadrant.

Exercise 7a. Transform the following into angles between 0° and 360°.

1. 390°	2. $-1°$	3. 627°	4. 425°
5. $-170°$	6. 563°	7. $-270°$	8. 1270°
9. $-180°$	10. 401°	11. $-401°$	12. $-760°$

7.2

We can now place any angle in one of the four quadrants. Each of the angles in the last three quadrants is equal in magnitude (size) to an angle in the first quadrant, although there may be a difference in sign. The angle is first drawn out roughly and then the acute angle between it and the horizontal line (in either direction) is calculated. This acute angle we shall call the basic angle. It will be an angle whose ratio sine, cosine or tangent is given in the standard four-figure tables.

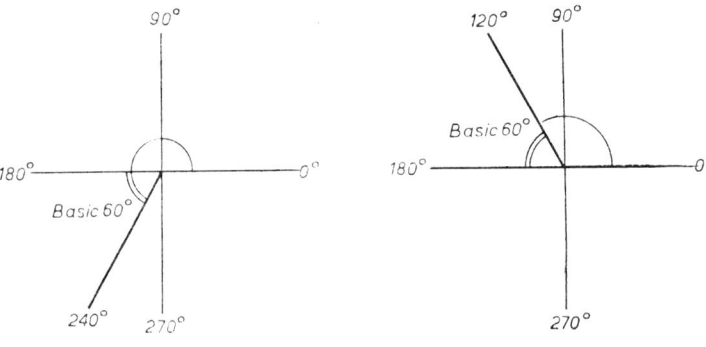

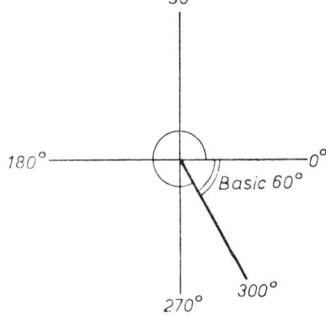

Thus 120° is equal to a basic angle of $180°-120°=60°$. 240° is equal to a basic angle of $240°-180°=60°$. 660° would first be transformed to 300° which is equal to basic angle $360°-300°=60°$.

Exercise 7b. Give the basic acute angles of:

1. 110°	2. 190°	3. 260°	4. 320°
5. 216°	6. 265°	7. 290°	8. 150°
9. 390°	10. 480°	11. −170°	12. −30°

7.3

The final step is to decide the sign and for this the ratios have to be considered separately. We start with the sine ratio. We shall consider the shape of the sine curve which is a graph of $y = \sin x$. We need only go from $x = 0°$ to $360°$, as it will then repeat itself in shape. It varies between 1 and $−1$. Angles along the base of the shaded area will fall

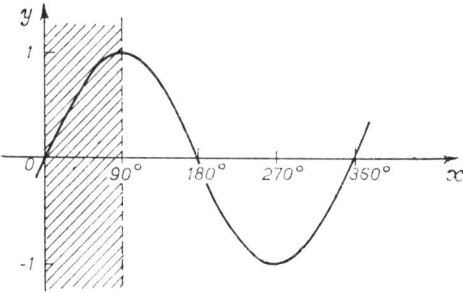

in the first quadrant, and for them $\sin x$ is above the axis and so it is positive. The second quadrant from 90° to 180° also has $\sin x$ positive. For the third and fourth quadrants $\sin x$ is below the axis and is negative.

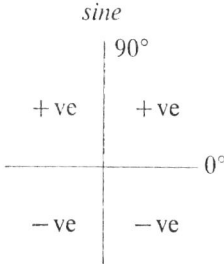

We can also note the following standard ratios –

$\sin 0° = 0$	$\sin 90° = 1$
$\sin 180° = 0$	$\sin 270° = −1$

As an example we can now find the value of $\sin 210°$. It is in the third quadrant with a basic angle of $210° − 180° = 30°$. For sine the third quadrant is negative, therefore $\sin 210° = -\sin 30°$. We use the tables to find $\sin 30°$, to give $\sin 210° = 0{\cdot}5000$.

$$\text{Again } \sin \frac{11\pi}{4} = \sin \frac{3\pi}{4} \text{ (subtracting } 360°, \text{ i.e. } 2\pi)$$

$$\sin \frac{3\pi}{4} = \sin 135°(\pi = 180°)$$

$$\sin 135° = +\sin 45° \text{ (2nd quadrant positive)}$$

$$\sin 45° = +0{\cdot}7071$$

$$\sin \frac{11\pi}{4} = = 0{\cdot}7071$$

Exercise 7c. Find the values of the following:

1. $\sin 100°$ 2. $\sin 240°$ 3. $\sin 300°$

4. $\sin 71°$ 5. $\sin 260°$ 6. $\sin 400°$

7. $\sin 270°$ 8. $\sin \dfrac{13\pi}{4}$ 9. $\sin(-70°)$

7.4

The cosine curve is the same shape as the sine curve but is displaced by 90°.

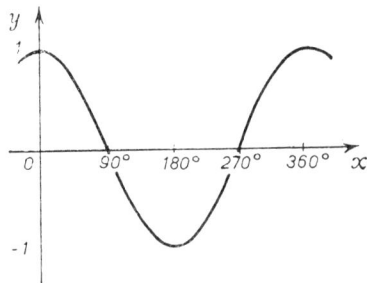

From this it is seen that the first quadrant angles have positive cosines the second and third have negative and the fourth positive.

cosine
90°

−ve	+ve
−ve	+ve

0°

The standard ratios are:

$$\cos 0° = 1 \qquad\qquad \cos 90° = 0$$
$$\cos 180° = -1 \qquad\qquad \cos 270° = 0$$

As an example find cosine 300°. It is in the fourth quadrant with a basic angle $360° - 300° = 60°$. For the fourth quadrant cosine is positive, therefore cosine $300° = +\cos 60° = +0{\cdot}5000$.

Exercise 7d. Find the values of the following:

1. $\cos 110°$ 2. $\cos 200°$ 3. $\cos 320°$

4. $\cos 10°$ 5. $\cos 600°$ 6. $\cos 270°$

7. $\cos \dfrac{15\pi}{4}$ 8. $\cos(-30°)$ 9. $\cos 707°$

7.5

The tangent curve has no scale marked on the y-axis as it reaches infinity (∞) for values of 90° and 270°. From this we see that tangent

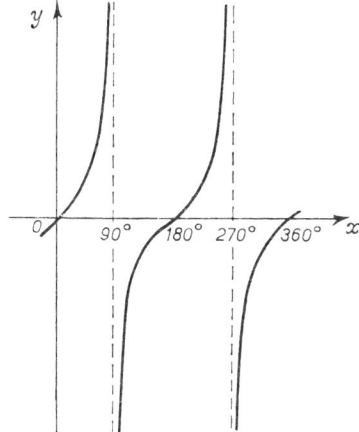

in the first quadrant is positive, in the second negative, third positive and fourth negative.

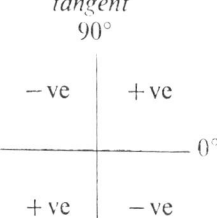

tangent
90°

The standard ratios are:

$$\tan 0° = 0 \qquad\qquad \tan 90° = \infty$$

$$\tan 180° = 0 \qquad\qquad \tan 270° = \infty$$

Thus $\tan 150°$ is in the second quadrant with basic angle

$$180° - 150° = 30°$$

For the second quadrant tangent is negative giving

$$\tan 150° = -\tan 30° = -0.5774$$

To find the secant of an angle of any magnitude the basic secant angle is found and the quadrant signs are the same as for its reciprocal ratio the cosine.

Cosecant has the quadrant signs of sine, and cotangent has the same as tangent.

Exercise 7e. Find the values of the following:

1. $\tan 100°$	2. $\tan 200°$	3. $\tan 310°$
4. $\tan 180°$	5. $\tan 620°$	6. $\tan 265°$
7. $\sec 330°$	8. $\operatorname{cosec} 120°$	9. $\cot 200°$

7.6

If we are given the signs of two ratios there will be only one quadrant in which the angle can fall as shown by studying the combined results.

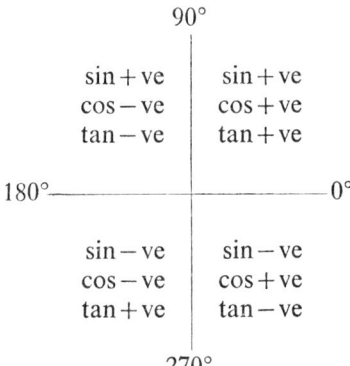

For example if we are given that $\sin x$ is equal to $-\frac{4}{5}$ and $\cos x$ is negative we can find any other ratio, such as $\tan x$. For the basic angle the opposite side and hypotenuse are in the ratio 4 to 5, i.e. if one is $4k$ the other is $5k$. Then by Pythagoras' theorem the adjacent side is given by $\sqrt{25k^2 - 16k^2}$, or $3k$. With a negative sine and negative cosine the angle is in the third quadrant and therefore has a positive tangent. The ratio opposite to adjacent is $4k$ to $3k$. Therefore $\tan x = +\frac{4}{3}$. If the angle is asked for, as well as a ratio, tables are used to find the basic angle.

Exercise 7f

1. If $\sin x = +\frac{4}{5}$ and $\cos x = -\frac{3}{5}$, find the value of $\tan x$.

2. If $\sin x = -\dfrac{\sqrt{3}}{2}$ and the tangent is positive, find the value of cosine x.

3. If $\cos x = -\dfrac{1}{\sqrt{2}}$ and the tangent is negative, find the value of $\sin x$ and $\operatorname{cosec} x$.

4. If $\cos x = \dfrac{-5}{13}$ and the sine is positive, find the value of $\tan x$.

5. If $\tan \theta = 1\cdot4$, find the possible values of θ between $0°$ and $360°$. Give the values of $\sin \theta$ for these angles.

6. If $\sec x = +\frac{13}{12}$ and the tangent is negative, find the value of $\sin x$.

7. If $\tan x = -\sqrt{3}$ and the sine is positive, find the value of $\cos x$.

8. If $\sin x = 0\cdot5$ and x lies between $0°$ and $360°$, find the possible values of x and the values of their cosines.

9. If x is an obtuse angle whose sine is $\frac{8}{17}$, find the value of $\cos x$ and $\cot x$.

7.7

Simple trigonometrical equations can be solved by the same methods as algebraic equations. For example, $6\cos^2 x + \cos x - 1 = 0$. This is a quadratic in $\cos x$ and goes into two brackets to give

$$(2\cos x + 1)(3\cos x - 1) = 0.$$

This gives $\cos x = -\frac{1}{2}$ or $+\frac{1}{3}$.

Taking the first result $\cos x = -\frac{1}{2}$, we find the basic angle from $\cos x = \frac{1}{2}$, which is a standard angle of section 3.5, i.e. $60°$. The negative sign for cosine gives possible angles in the second or third quadrant, i.e. $120°$ or $240°$ $(180° - 60°$ or $180° + 60°)$.

Taking the second result we find the basic angle by looking up $\cos x = 0\cdot3333$ in the tables. This gives $70° 32'$. The positive sign for cosine gives two angles, one in the first and the other in the fourth quadrant, i.e. $70° 32'$ and $289° 28'$. The full solution in the range $0°$ to $360°$ is $x = 70° 32'$, $120°$, $240°$ or $289° 28'$. Sometimes solutions are asked for in the range $-180°$ to $180°$, in which case $360°$ would be subtracted from each of the final two angles.

Sometimes the relationship between ratios of section 3.4 have to be used in solutions. For example, $2\sin x = \tan x$, i.e. $2\sin x = \dfrac{\sin x}{\cos x}$

$$\therefore\ 2\sin x . \cos x = \sin x$$

$$\sin x(2\cos x - 1) = 0$$

$$\therefore\ \sin x = 0, \quad \text{i.e. } x = 0°, \ 180° \text{ or } 360°,$$

$$\text{or } \cos x = +\tfrac{1}{2}, \text{ i.e. } x = 60° \text{ or } 300°.$$

The values of x which satisfy the equation in the range $0°$ to $360°$ are $x = 0°, 60°, 180°, 300°$ or $360°$.

Exercise 7g. Find the values of x between $0°$ and $360°$ which satisfy the following equations:

1. $(2\sin x - 1)(3\cos x + 2) = 0$
2. $6\sin^2 x + \sin x - 1 = 0$
3. $2\cos^2 x = \cos x$
4. $\tan^2 x + \dfrac{5}{\cot x} + 6 = 0$
5. $\sec^2 x - 5\tan x + 5 = 0$
6. $2\sin^2 x = 1$
7. $3\sin x = \cos x$
8. $4\cos x . \sin 2x = \cos x . \cos 2x$
9. $6\cos^2 x - 5\cos x + 1 = 0$
10. $\sin^2 x = 2 - 2\cos x$
11. $\sin 2x = 3\cos 2x$
12. $\cos^2 x - \sin^2 x = \sin x$

8. Geometry of Distances, Angles, and Areas

8.1

To find the distance between the points $P_1(x_1, y_1)$ and $P_2(x_2, y_2)$ it is best to construct the lines $x = x_1$, $x = x_2$, $y = y_1$, and $y = y_2$.

Since x_1 is a co-ordinate the distance of the points $(x_1, 0)$ from the origin is also x_1, and a similar result applies to the other points. From the diagram it can be seen that $P_1 N = x_2 - x_1$ and $P_2 N = y_2 - y_1$.

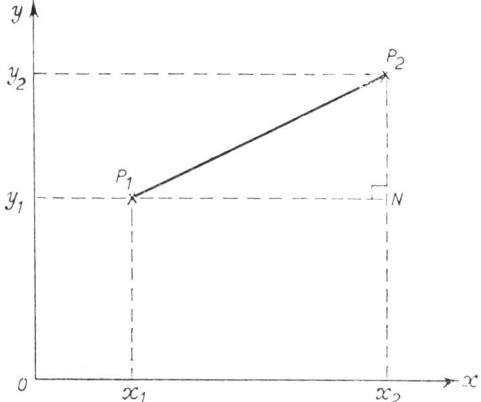

If d is the distance $P_1 P_2$ then by Pythagoras' theorem

$$d^2 = (x_2 - x_1)^2 + (y_2 - y_1)^2$$

This is often the most useful form of the equation in problems, but in simple calculations the positive root is taken giving:

$$d = +\sqrt{(x_2 - x_1)^2 + (y_2 - y_1)^2}$$

From this we can obtain the distance of $P_1(x_1, y_1)$ from the origin. In this case P_2 is $(0, 0)$ giving $OP_1 = \sqrt{x_1^2 + y_1^2}$.

Let us find the distance between the points $(-2, 3)$ and $(4, -5)$. It does not matter which point is made P_1, as each difference is squared and will give the same result. So

$$(x_1, y_1) \text{ is } (-2, 3) \text{ and } (x_2, y_2) \text{ is } (4, -5)$$

then
$$d^2 = [4 - (-2)]^2 + [-5 - 3]^2$$
$$= 6^2 + (-8)^2$$
$$= 36 + 64$$
$$= 100$$
$$\therefore d = +10 \text{ units}$$

Exercise 8a. Find, giving the answers to three significant figures if square-root tables are used, the distances between:

1. $(0, -3)$ and $(6, 5)$
2. $(-2, -2)$ and $(2, 2)$
3. $(-1, 7)$ and $(11, 2)$
4. $(3, 2)$ and $(4, 6)$
5. $(-1, -4)$ and $(-3, -2)$
6. $(5, -1)$ and the origin

8.2

To find the co-ordinates of the point $P(\bar{x},\bar{y})$ which divides the line joining the points $P_1(x_1,y_1)$ and $P_2(x_2,y_2)$ internally in the ratio $k_1:k_2$ we again construct the x and y lines.

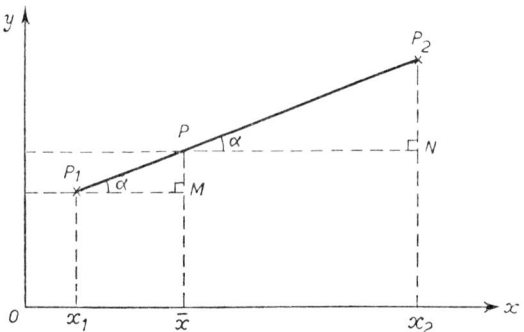

In $\triangle P_1 PM$ the line $P_1 M = \bar{x} - x_1$

$$\therefore \cos\alpha = \frac{\bar{x} - x_1}{P_1 P} \tag{1}$$

In $\triangle PP_2 M$ the line $PN = x_2 - \bar{x}$

$$\therefore \cos\alpha = \frac{x_2 - \bar{x}}{PP_2} \tag{2}$$

From equations (1) and (2):

$$\frac{\bar{x} - x_1}{P_1 P} = \frac{x_2 - \bar{x}}{PP_2}$$

$$\therefore \frac{\bar{x} - x_1}{x_2 - \bar{x}} = \frac{P_1 P}{PP_2} = \frac{k_1}{k_2}$$

$$k_2\bar{x} - k_2 x_1 = k_1 x_2 - k_1\bar{x}$$

$$k_2\bar{x} + k_1\bar{x} = k_1 x_2 + k_2 x_1$$

$$\bar{x} = \frac{k_1 x_2 + k_2 x_1}{k_1 + k_2}$$

Since $\sin\alpha = \dfrac{\bar{y} - y_1}{P_1 P}$ and also $\dfrac{y_2 - \bar{y}}{PP_2}$ we obtain a similar result

$$\bar{y} = \frac{k_1 y_2 + k_2 y_1}{k_1 + k_2}$$

A special case is the mid-point of $P_1 P_2$, when $k_1 = k_2$. Then

$$\bar{x} = \frac{k_1 x_2 + k_1 x_1}{2k_1} = \frac{x_2 + x_1}{2}$$

Thus the mid-point is given by

$$\left(\frac{x_2 + x_1}{2}, \frac{y_2 + y_1}{2}\right)$$

e.g. to find the point which divides the line joining $(2, 3)$ and $(5, -2)$ internally in the ratio $1:2$.

$$\bar{x} = \frac{1 \times 5 + 2 \times 2}{1 + 2}; \quad \bar{y} = \frac{1 \times (-2) + 2 \times 3}{1 + 2}$$

$$\bar{x} = \frac{9}{3} = 3; \qquad \bar{y} = \frac{4}{3}$$

The required point is $(3, \frac{4}{3})$.

8.3

For the point $P(\bar{x}, \bar{y})$, which divides the line joining P_1 and P_2 externally, the ratio $k_1 : k_2$ has to be negative. Thus k_1 can be given a negative sign.

When the external point P is beyond P_2 k_1 automatically becomes numerically larger than k_2 as shown below.

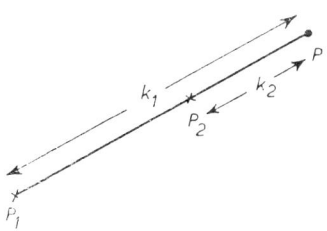

In the same way when it is beyond P_1, k_2 must be numerically larger than k_1.

For example to find the point which divides the line joining the points $(2, 3)$ and $(5, -2)$ externally in the ratio $1:2$ we use the formulae of the previous section with $k_1 = -1$ and $k_2 = +2$.

$$\bar{x} = \frac{(-1) \times 5 + 2 \times 2}{(-1) + 2}; \quad \bar{y} = \frac{(-1) \times (-2) + 2 \times 3}{(-1) + 2}$$

$$\bar{x} = \frac{-1}{1} = -1; \qquad \bar{y} = \frac{8}{1} = 8$$

The required point is $(-1, 8)$.

Exercise 8b

1. Find the point which divides the line joining the points $(2,3)$ and $(5,6)$ internally in the ratio of $2:1$.

2. Find the point which divides the line joining the points $(2,3)$ and $(5,6)$ externally in the ratio of $2:1$.

3. Find the point which divides the line joining the points $(2,3)$ and $(5,6)$ externally in the ratio of $1:2$.

4. Find the mid-point of the line joining $(2,3)$ and $(5,6)$.

5. Find the point which divides the line joining the points $(2,3)$ and $(5,6)$ internally in the ratio of $1:2$.

6. Find the point which divides the line joining the points $(0,-4)$ and $(-3,5)$ internally in the ratio of $3:4$.

7. Find the point which divides the line joining the points $(2,-2)$ and $(-5,7)$ in the ratio $-3:5$.

8. Find the mid-point of the line joining $(-3,-2)$ and $(4,7)$.

8.4

The angle θ between the two straight lines $y = m_1 x + c_1$ and $y = m_2 x + c_2$ is found by considering the slopes m_1 and m_2.

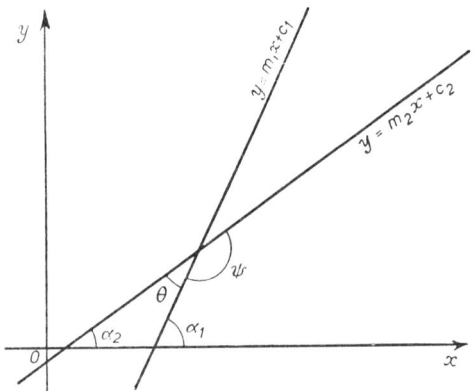

The slope m_1 is by definition $\tan \alpha_1$ and the slope m_2 is $\tan \alpha_2$. The external angle α_1 of the triangle formed by the two lines and x-axis is equal to the sum of the opposite internal angles $\theta + \alpha_2$, i.e.

$$\theta = \alpha_1 - \alpha_2$$

$$\tan \theta = \tan(\alpha_1 - \alpha_2)$$

In section 14.2 it will be shown that

$$\tan(\alpha_1 - \alpha_2) = \frac{\tan \alpha_1 - \tan \alpha_2}{1 + \tan \alpha_1 . \tan \alpha_2}$$

Substituting the values $\tan \alpha_1 = m_1$ and $\tan \alpha_2 = m_2$ we obtain

$$\tan \theta = \frac{m_1 - m_2}{1 + m_1 . m_2}.$$

If the value of the right-hand side is negative it represents the tangent of the supplementary angle ψ. This is because the first slope taken will have been the less steep of the two.

There are two important special cases. Firstly, if the two lines are parallel the angle between them is zero, i.e. $\tan \theta = 0$

$$\therefore \frac{m_1 - m_2}{1 + m_1 . m_2} = 0; \quad m_1 - m_2 = 0 \quad \text{giving } m_1 = m_2.$$

Parallel lines have the same slope.

Secondly, two perpendicular lines have an angle of 90° between them, i.e. $\tan \theta$ infinity.

$$\therefore \frac{m_1 - m_2}{1 + m_1 . m_2} = \infty$$

For a fraction to equal infinity its denominator must be zero.

$$\therefore 1 + m_1 . m_2 = 0, \quad \text{or } m_1 = -\frac{1}{m_2}$$

The slope of a line is minus the reciprocal of the slope of any line perpendicular to it.

To find the acute angle between the lines $2y = x + 7$ and $3x + 2y = 7$ we first of all find the two slopes:

$$2y = x + 7 \quad \therefore y = \tfrac{1}{2}.x + \tfrac{1}{2} \quad \text{i.e. } m_1 = \tfrac{1}{2}$$
$$3x + 2y = 7 \quad \therefore y = -\tfrac{3}{2}x + \tfrac{7}{2} \quad \text{i.e. } m_2 = -\tfrac{3}{2}$$
$$\tan \theta = \frac{\tfrac{1}{2} - (-\tfrac{3}{2})}{1 + \tfrac{1}{2}.(-\tfrac{3}{2})}$$
$$= \frac{2}{\tfrac{1}{4}} = 8.$$
$$\theta = 82° 52'$$

Exercise 8c

1. Find the slope of any line parallel to the line $5y + 4x = 6$.

2. Find the slope of any line perpendicular to the line $1 + 2x = 3y$.

D

3. Find the acute angle between the lines $y = 2x + 3$ and $y = 4x + 7$.
4. Find the acute angle between the lines $2y = 3x + 6$ and $2y + 3 = 5x$.
5. Find the obtuse angle between the lines $4y + 3 = 2x$ and $y = 4x + 3$.
6. Find the angle between the lines $2(x + y) = 3$ and $y = x + 5$.

8.5

The area of a triangle, given the vertices (x_1, y_1); (x_2, y_2) and (x_3, y_3), can be stated in the form of a matrix.

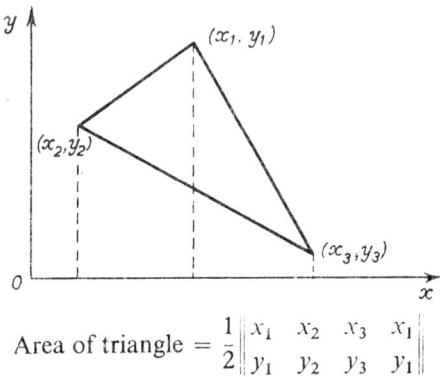

$$\text{Area of triangle} = \frac{1}{2} \begin{Vmatrix} x_1 & x_2 & x_3 & x_1 \\ y_1 & y_2 & y_3 & y_1 \end{Vmatrix}$$

To multiply out we obtain three positive pairs by multiplying diagonally from the x's, i.e. $+x_1 y_2$; $+x_2 y_3$; $+x_3 y_1$. Then three negative pairs from the y's, i.e. $-y_1 x_2$; $-y_2 x_3$; $-y_3 x_1$. The signs of the actual x and y values also have to be taken into account. If the answer is negative it is because the order of vertices has been taken in a clockwise direction, but it is numerically correct. The area formula as multiplied out can be proved by considering the areas of the three trapeziums formed by the construction lines.

To find the area of a triangle with vertices $(3, 4)$, $(-5, 2)$, and $(6, -1)$ we set out the matrix

$$\text{Area} = \frac{1}{2} \begin{Vmatrix} 3 & -5 & 6 & 3 \\ 4 & 2 & -1 & 4 \end{Vmatrix}$$

$$+\text{ve} \qquad\qquad -\text{ve}$$

$$\tfrac{1}{2}\{3.2 - 4.(-5) + (-5)(-1) - 2.6 + 6.4 - (-1).3\}$$

Here the negative signs inside brackets are due to the actual numbers, and those outside to the matrix.

$$\tfrac{1}{2}\{6 + 20 + 5 - 12 + 24 + 3\} = \tfrac{1}{2}(46) = 23 \text{ square units}$$

If the area is zero the three vertices are actually three points in a straight line.

8.6

There is a useful result which applies to the vertices of a parallelogram, which of course includes rectangles, diamonds, squares, etc. If the vertices in order are (x_1, y_1), (x_2, y_2), (x_3, y_3), and (x_4, y_4), then

$$x_1 + x_3 = x_2 + x_4 \quad \text{and} \quad y_1 + y_3 = y_2 + y_4$$

These are easily proved as the diagonals bisect each other. If there is any doubt about the position of the vertices they should be plotted.

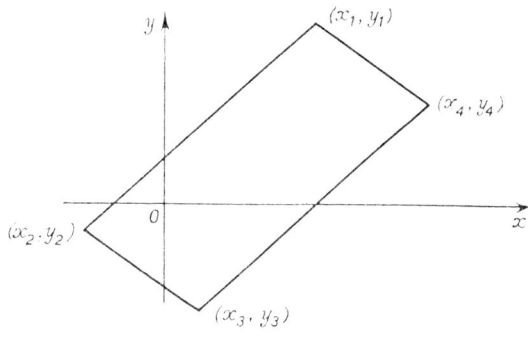

For example find the fourth vertex of a parallelogram with vertices, taken in order: $(7, 8)$, $(-1, 3)$, and $(0, -4)$.

$$x_1 + x_3 = x_2 + x_4 \qquad y_1 + y_3 = y_2 + y_4$$
$$7 + 0 = -1 + x_4 \qquad 8 - 4 = 3 + y_4$$
$$8 = x_4 \qquad 1 = y_4$$

The vertex is $(8, 1)$.

Exercise 8d

1. Find the area of the triangle with the vertices $(2, 1)$, $(5, 3)$ and $(4, 6)$.

2. Find the area of the triangle with vertices $(2, 1)$, $(-1, 2)$, and $(-3, -2)$.

3. What is the area of the triangle with vertices $(-3, -1)$, $(-2, 3)$, and $(3, -4)$?

4. A triangle has an area of 20 square units and two vertices are $(3, 4)$ and $(2, 7)$. What are the possible third vertices if they lie on the x axis?

5. What is the area of the triangle with vertices $(0, 1)$, $(3, 5)$, and $(-3, -3)$? How do you interpret this?

6. If $(0, 2)$, $(-4, 5)$, and $(2, -2)$ are three corners of a parallelogram, find the co-ordinates of the three points which can complete the parallelogram.

9. Function of a Function

9.1

If $y = u^7$ and $u^2 = x$ there is no need to eliminate u to obtain $\dfrac{dy}{dx}$. y is a function of u which in turn is a function of x. To find $\dfrac{dy}{dx}$ we can introduce another variable to split it up into $\dfrac{dy}{du} \cdot \dfrac{du}{dx}$:

$$y = u^7 \qquad\qquad\qquad u = x^{1/2}$$

$$\frac{dy}{du} = 7u^6 \qquad\qquad\qquad \frac{du}{dx} = \tfrac{1}{2}x^{-1/2}$$

$$\therefore \frac{dy}{dx} = \frac{dy}{du} \cdot \frac{du}{dx} = 7u^6 \cdot \frac{1}{2\sqrt{x}} = \frac{7}{2} \cdot \frac{u^6}{\sqrt{x}} = \frac{7x^3}{2\sqrt{x}} = \frac{7}{2} \cdot x^{5/2}$$

(u^6 is replaced by x^3 from the original equation $u^2 = x$).

9.2

Another useful result, which applies to any first derivative, is that

$$\frac{dy}{dx} = \frac{1}{\dfrac{dx}{dy}}$$

Thus if $y^5 = x$, then $\dfrac{dx}{dy} = 5y^4$.

$$\therefore \frac{dy}{dx} = \frac{1}{5y^4} = \frac{1}{5x^{4/5}}$$

by substitution from the original equation.

9.3

The rule of section 9.1 can be extended to any number of functions. For example if $y = u^7$, $u = v^2$, and $v = x^3$, then

$$\frac{dy}{du} = 7u^6; \quad \frac{du}{dv} = 2v \quad \text{and} \quad \frac{dv}{dx} = 3x^2$$

$$\frac{dy}{dx} = \frac{dy}{du}\cdot\frac{du}{dv}\cdot\frac{dv}{dx} = 7u^6.2v.3x^2$$

$$= 42v^{12}.v.x^2$$

$$= 42.x^{36}.x^3.x^2$$

$$= 42x^{41}$$

9.4

When both sides of the equation involve powers we can extend the first rule.

$$y^3 = x^4$$

$$\therefore \frac{d(y^3)}{dx} = 4x^3$$

and making it a function of a function

$$\frac{d(y^3)}{dy}\cdot\frac{dy}{dx} = 4x^3$$

But the differential of y^3 with respect to y is $3y^2$

$$\therefore 3y^2.\frac{dy}{dx} = 4x^3$$

If necessary the y^2 can be substituted using the original equation. This process can be done in one step, e.g.

$$y^7 = 3x^2+4x+7$$

$$7y^6.\frac{dy}{dx} = 6x+4$$

Exercise 9a. Find $\frac{dy}{dx}$ in terms of x for:

1. $y = u^6; u = x^4$
2. $y = z^3; z^5 = x$
3. $y = 3t^4; t = \frac{7}{u}; u^2 = x$

Without substituting for y's in the answer, find $\dfrac{dy}{dx}$ for:

4. $y^7 = x^5$
5. $2y^5 = x(1+x)$
6. $y^2 = (1+x)(3+x^2)$
7. $y^7 = x$
8. $2y^3 + 7 = x$

9.5

By far the most important use of the function of a function rule is to differentiate powers of functions of x. As an example we take $y = (2x^2+3)^4$. It would be very tedious, and quite unnecessary, to multiply out, differentiate and then factorise again. We replace $2x^2+3$ by a new function, which we shall call u.

Then
$$y = u^4 \quad \text{and} \quad u = 2x^2 + 3$$

$$\frac{dy}{du} = 4u^3 \quad \text{and} \quad \frac{du}{dx} = 4x$$

$$\therefore \frac{dy}{dx} = \frac{dy}{du} \cdot \frac{du}{dx} = 4u^3 \cdot 4x$$

Finally, substituting for u:

$$\frac{dy}{dx} = 16x \cdot (2x^2+3)^3$$

Exercise 9b. Differentiate with respect to x:

1. $(x+7)^3$ 2. $(2x+1)^5$ 3. $(2x^2+7)^6$ 4. $(x^2+1)^{1/2}$

5. $\sqrt[3]{x^2+1}$ 6. $\sqrt[4]{2x^5+3x}$ 7. $(x^2+1)^{-2}$ 8. $\dfrac{1}{(x^2+1)^3}$

9. $\dfrac{1}{(2x^5+3x)^4}$ 10. $(6x+5)^2 + (3x+2)$

11. $(1+x^2)^3 + (1-x)^2$ 12. $(2x+3) + \sqrt{x^2+1}$

9.6

Since this type of differentiation has to be used so often we must look carefully at the answers so that we can write them down in one step. $y = (2x^2+3)^4$ is first differentiated as if the bracket is X, giving $4X^3$, i.e. $4(2x^2+3)^3$, and to compensate for having a bracket instead of X we multiply by the differential of the bracket, $4x$.

$$\therefore \frac{dy}{dx} = 4(2x^2+3)^3 \cdot 4x = 16x(2x^2+3)^3$$

Taking a further example of the product type:

$$y = (2x+1)^2.(3x+2)$$

By the product rule the derivative is given by multiplying the first part $(2x+1)^2$ by the differential of the second 3. To this add the multiple of the second part $(3x+2)$ and the differential of the first $2(2x+1)^1.2$, this part being a function of a function.

$$\frac{dy}{dx} = (2x+1)^2.3+(3x+2).4(2x+1)$$
$$= (2x+1)\{6x+3+12x+8\}$$
$$= (2x+1)(18x+11)$$

Exercise 9c. Repeat the differentials of 1–12 of Exercise 9b, doing them in one step.

13. $(1+x^2)^2.(3+4x)$ 14. $(2+x)^2.(3-x^2)^2$

---

10. Solutions of Triangles; Area

10.1

To solve a triangle we have to find any unknown sides or angles. When we know two angles we can always find the third from the fact that they all add up to 180°.

When we are given one side and two angles the triangle is solved using the sine rule. This rule can also be used when we are given two sides and one angle which is not included if some indication is given as to whether one of the other angles is obtuse or not.

To prove the sine rule we construct an altitude CN.

In $\triangle ACN$:

$$CN = b.\sin A$$

In $\triangle BCN$:

$$CN = a.\sin B$$
$$\therefore a.\sin B = b.\sin A$$

i.e. $\dfrac{a}{\sin A} = \dfrac{b}{\sin B}$

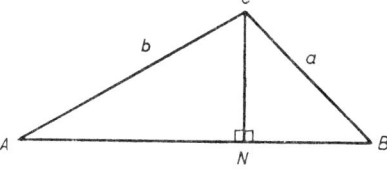

By using a similar method after constructing altitude AM we can show

$$\frac{b}{\sin B} = \frac{c}{\sin C}, \text{ where } C \text{ may be obtuse.}$$

Thus the sine rule gives us

$$\frac{a}{\sin A} = \frac{b}{\sin B} = \frac{c}{\sin C}$$

Taking the first example, with two angles and one side, $B = 70°$, $C = 80°$, and $a = 5 \cdot 00 \, \text{m}$.

$$\text{Angle } A \ = \ 180° - (B + C) \ = \ 180° - 150° \ = \ 30°$$

$$\therefore \ \frac{5}{\sin 30°} = \frac{b}{\sin 70°} \quad \text{i.e. } b = \frac{5 \cdot \sin 70°}{\sin 30°} = \frac{5 \times 0 \cdot 9397}{0 \cdot 5000}$$

$$b = 9 \cdot 40 \, \text{m}$$

Also

$$\frac{5}{\sin 30°} = \frac{c}{\sin 80°} \quad \text{i.e. } c = \frac{5 \cdot \sin 80°}{\sin 30°} = \frac{5 \times 0 \cdot 9848}{0 \cdot 5000}$$

$$c = 9 \cdot 85 \, \text{m}$$

If one of the angles is obtuse we use the result of section 7.3, e.g. $\sin 110° = +\sin(180 - 110)° = +\sin 70°$.

It is this result giving sine positive in the first two quadrants that leads to ambiguity with two sides and the non-included angle given. Of course if the angle given is obtuse the other two unknown angles must be acute, but we shall take a case where the known angle is acute.

Solve the triangle with $a = 5 \cdot 00$ cm., $b = 6 \cdot 00$ cm. and $A = 30°$.

$$\frac{a}{\sin A} = \frac{b}{\sin B} \quad \therefore \ \sin B = \frac{b \sin A}{a} = \frac{6 \cdot 00 \times 0 \cdot 5000}{5 \cdot 00} = 0 \cdot 6000$$

From section 7.3 we see that $\sin B = 0 \cdot 6000$ can have B in the first or second quadrant.

From the tables $B_1 = 36° \, 52'$. Also $B_2 = 180 - 36° \, 52' = 143° \, 08'$.

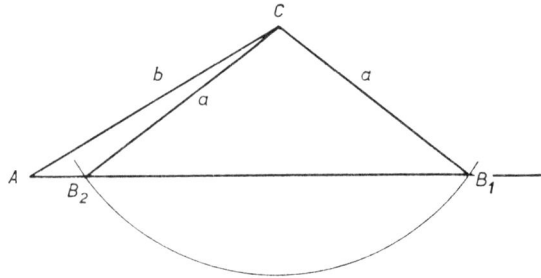

The diagram shows the possible triangles ACB_1 and ACB_2, so some definite guide must be given in the question. Since angle C is the third angle it can be found and then side c.

Let us now consider the case where b is less than a. If a is again 5·00 cm. and $A = 30°$ but b is only 4·00 cm., then

$$\sin B = \frac{b \sin A}{a} = \frac{4·00 \times 0·5000}{5·00} = 0·4000$$

Side b is less than side a and therefore angle B must be less than angle $A(30°)$. The only possible angle is the basic one, $23° 35'$.

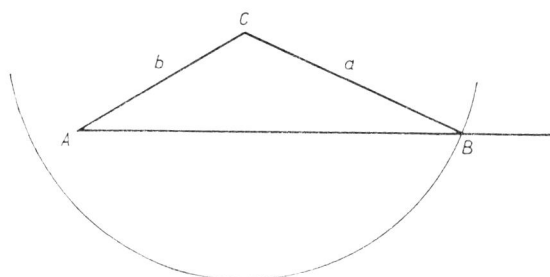

The diagram shows only one position of B, as A has to be $30°$. Again, angle C is found and then side c.

Exercise 10a. Solve the following triangles:

1. $a = 5·00$ cm., $b = 7·00$ cm., $A = 30°$, and B is acute.
2. $a = 5·00$ cm., $b = 7·00$ cm., $A = 30°$, and C is acute.
3. $a = 5·00$ cm., $b = 3·50$ cm., $A = 30°$.
4. $c = 7·21$ m, $A = 40°$, $B = 60°$.
5. $b = 7·51$ m, $c = 8·91$ m, $B = 50°$, and C obtuse.
6. $b = 4·54$ m, $A = 40°$, $B = 110°$.
7. $b = 7·51$ m, $c = 8·91$ m, $B = 50°$, and C acute.
8. $c = 9·23$ m, $a = 12·31$ m, $A = 70°$.

10.2

When we are given three sides or two sides and the included angle we start the solution using the cosine rule. Since second quadrant cosines are negative, as shown in section 7.4, there is only one possible answer in each case for the cosine rule. We should expect this as $(S.S.S)$ and $(S.A.S)$ are conditions for congruence, the same applying to $(A.S.A)$ in the previous section. To prove the cosine rule we construct altitude CN and let $AN = x$.

$$\therefore BN = c - x$$

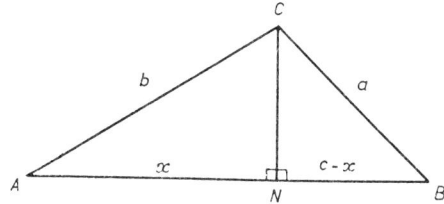

By Pythagoras' theorem:

In $\triangle ACN$: $\qquad\qquad CN^2 = b^2 - x^2$

In $\triangle BCN$: $\qquad\qquad CN^2 = a^2 - (c - x)^2$

$$\therefore a^2 - c^2 + 2cx - x^2 = b^2 - x^2$$

$$a^2 = b^2 + c^2 - 2cx$$

but from $\triangle ACN$: $x = b\cos A$.

$$\therefore a^2 = b^2 + c^2 - 2bc \cos A$$

Similarly $\qquad\qquad b^2 = c^2 + a^2 - 2ca \cos B$

$$c^2 = a^2 + b^2 - 2ab \cos C, \text{ where } C \text{ may be obtuse.}$$

When we are given three sides we use the cosine rule to find the largest angle, i.e. that opposite the largest side. If $a = 3$ cm., $b = 4$ cm. and $c = 6$ cm., we find angle C

$$36 = 9 + 16 - 24 . \cos C$$

$$24 \cos C = 25 - 36 = -11$$

$$\cos C = -\tfrac{11}{24} = -0.4583$$

$$C = 180 - 62° 43' = 117° 17'$$

The second angle (B or A) can be found using the sine rule. Since only the largest angle can be obtuse, the second angle must be acute.

$$\sin A = \frac{a \sin C}{c} = \frac{3 . \sin 117° 17'}{6} = \frac{\sin 62° 43'}{2}$$

$$\therefore \sin A = \frac{+0.8888}{2} = 0.4444$$

$$A = 26° 23'$$

$B = 180° - 117° 17' - 26° 23' = 36° 20'$, which solves the triangle.

Given two sides and the included angle we have to find the side opposite the angle. For example, $c = 3$ cm., $a = 4$ cm., and $B = 120°$.

$$b^2 = 9 + 16 - 24 \cos 120°$$

$$= 25 - 24(-\cos 60°)$$

$$= 25 + 24 . \tfrac{1}{2} = 37$$

$$b = 6.083 \text{ cm.}$$

Since B is the largest angle, either C or A can be found using the sine rule. If the middle angle had been given we should have found the smallest next, and vice versa. This, then, leaves the largest angle as the third angle and it is automatically fixed, by the angles of a triangle property, as acute or obtuse.

Exercise 10b. Solve the following triangles:

1. $a = 5$ cm., $b = 12$ cm., $c = 15$ cm.
2. $a = 5$ cm., $b = 12$ cm., $c = 12 \cdot 5$ cm.
3. $b = 5$ cm., $c = 4$ cm., $A = 60°$.
4. $b = 10$ cm., $c = 4$ cm., $A = 20°$.
5. $a = 10$ cm., $b = 5$ cm., $c = 6$ cm.
6. $a = 7$ cm., $b = 8$ cm., $C = 150°$.
7. $a = 7$ cm., $b = 8$ cm., $C = 30°$.
8. $a = 7 \cdot 8$ m, $b = 8 \cdot 0$ m, $c = 12 \cdot 2$ m.

10.3

To find the area of triangle ABC we again construct the altitude CN.

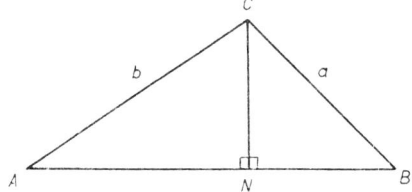

The   Area of the triangle $= \frac{1}{2} . AB . CN$

From $\triangle ACN$ Area $= \frac{1}{2} c . b . \sin A$

From $\triangle BCN$ Area $= \frac{1}{2} c . a . \sin B$

And, by a similar method, area $= \frac{1}{2} . a . b . \sin C$, where C may be obtuse. Thus

Area of a triangle $= \frac{1}{2} ab . \sin C$ or $\frac{1}{2} b . c . \sin A$ or $\frac{1}{2} . c . a . \sin B$

For example let us find the area outside a regular hexagon of side 2 m but inside its circum-circle.

The regular hexagon is formed by six equilateral triangles.

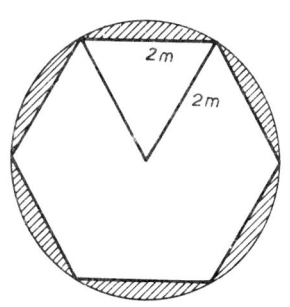

Area of one triangle $= \dfrac{1}{2} . 2 . 2 . \sin 60°$

$$= 2 . \frac{\sqrt{3}}{2} = \sqrt{3}$$

Area of hexagon $= 6 . \sqrt{3}$

Area of circle $= \pi(2)^2 = 4\pi$

$\therefore$ Shaded area $= 4\pi - 6 . \sqrt{3}$ m^2

10.4

If three sides of a triangle (a, b, and c) are given, the

$$\text{perimeter} = a+b+c$$

The semi-perimeter $s = \dfrac{a+b+c}{2}$.

By Hero's formula the area of the triangle equals

$$\sqrt{s(s-a)(s-b)(s-c)}$$

To find the area of a triangle of sides $a = 7$ cm., $b = 6$ cm., and $c = 5$ cm. we firstly find the semi-perimeter

$$s = \frac{7+6+5}{2} = 9\,\text{cm.}$$

$$\therefore (s-a) = 2\,\text{cm.}; (s-b) = 3\,\text{cm., and } (s-c) = 4\,\text{cm.}$$

(as a check these three terms when added must equal s).

$$\therefore \text{area} = \sqrt{9.2.3.4}$$

$$= 6\sqrt{6}\ \text{cm}^2$$

Exercise 10c

1. Find the area of a triangle of sides 3 cm., 4 cm., and 5 cm. by Hero's formula.
2. Find the area of a triangle with $a = 21.4$ cm., $b = 10.3$ cm., and $C = 40°$.
3. Find the area outside a regular pentagon but inside its circum-circle of radius 3 cm.
4. Find the area inside a regular pentagon but outside its inscribed
· circle of radius 3 cm.
5. Find the area of a triangle of sides 5 cm., 12 cm., and 16 cm.
6. The area of a triangle is 23 cm^2, and two of its sides are 9 cm and 10 cm. What is the angle included by these sides?
7. What is the area of a triangle sides 7 cm., 8 cm., and 9 cm.?
8. What is the area of a triangle sides 7 cm., 24 cm., and 25 cm.?
9. Triangle ABC has a circum-circle of radius R.
   Prove that $\dfrac{a}{\sin A} = 2R$.

11. Applications of Differentiation, Maxima, and Minima

11.1

We have found the limiting value of the rate of change $\frac{\delta y}{\delta x}$ when δx has become infinitesimal. δx is said to have tended to zero ($\delta x \to 0$).

Thus the first derivative $\frac{dy}{dx} = \underset{\delta x \to 0}{\text{Lt}}\ \frac{\delta y}{\delta x}$. The first derivative can also be written $f'(x)$ or $F'(x)$, if the original expression is represented by $f(x)$ or $F(x)$. We have seen that the gradient of a curve is the geometrical interpretation of the limit of $\frac{\delta y}{\delta x}$.

11.2

A very special limiting value is that for a rate of change of distance with time. If a small change in distance δs takes place in time δt the average velocity is $\frac{\delta s}{\delta t}$. In the limit as $\delta t \to 0$ we obtain $\frac{ds}{dt}$ which is the velocity at any instant. Since rates of change with time as the independent variable are so important in mechanics a special form is given to the first derivative, this is $\dot{s}$. Thus when s represents distance, t time and v velocity,

$$v = \frac{ds}{dt} \quad \text{or } \dot{s}$$

Suppose $s = ut + \frac{1}{2}at^2$, where u and a are constants then:

$$v = \dot{s} = u + \frac{1}{2}.a.2t$$

by the rules for finding first derivates. The velocity after 5 seconds would be $v = u + 5a$ units of distance per second.

11.3

Acceleration is defined as the rate of change of velocity. For a small change δv in time δt the average acceleration is $\frac{\delta v}{\delta t}$. The instantaneous acceleration is therefore the limit of $\frac{\delta v}{\delta t}$, when $\delta t \to 0$, i.e. $\frac{dv}{dt}$ or $\dot{v}$.

But we have already shown that $v = \dfrac{ds}{dt}$, therefore acceleration is

$$\frac{d\left(\dfrac{ds}{dt}\right)}{dt}$$

which is $\dfrac{d^2s}{dt^2}$, the second derivative of s. Using our special form this is $\ddot{s}$

$$\text{Acceleration} = \dot{v} = \ddot{s}$$

In the last section the example when differentiated with respect to time gave $v = u + at$. If it is differentiated again with respect to time we obtain acceleration $= \dot{v} = a$. But a was given as a constant so the acceleration is constant in this example.

11.4

If velocity is given in terms of s we can only differentiate with respect to distance. Let us consider the acceleration in the form $\dfrac{dv}{dt}$. By our function-of-a-function rule of section 9.1 this can be written as $\dfrac{dv}{ds} \cdot \dfrac{ds}{dt}$.

But by section 11.2 $\dfrac{ds}{dt} = v$. The acceleration is thus $v \cdot \dfrac{dv}{ds}$.

Let us find the acceleration if $v = s + s^2$.

$v = s + s^2$, where s is the distance in metres.

$$\therefore \frac{dv}{ds} = 1 + 2s$$

$$\text{Acceleration} = v \cdot \frac{dv}{ds} = (s + s^2)(1 + 2s)$$

Thus after 2 m:

$$s = 2 \text{ m}$$

$$v = 2 + 4 = 6 \text{ m/s}$$

11.5 $a = 6(1 + 4) = 30 \text{ m/s}^2$

In Chapter 2 we considered how a small increment δx in x caused a change δy in y. It was seen that the difference between $\dfrac{\delta y}{\delta x}$ and $\dfrac{dy}{dx}$ was only terms involving δx or powers of δx. Since δx is small

$$\frac{\delta y}{\delta x} \simeq \frac{dy}{dx}$$

$$\therefore \delta y = \frac{dy}{dx} \cdot \delta x$$

to an accuracy within 1% of the exact answer, because δx must always be less than 1% of x if it is a small increment.

If a cube is made of side 10 m within an accuracy of 10 mm, by how much can the volume vary from the exact 1000 m³?

Let volume be v and side l.

$$v = l^3$$

and change in length $\delta l = 0.01$ m

$$\therefore \frac{dv}{dl} = 3l^2$$

$\therefore$ maximum error in volume $\delta v = \frac{dv}{dl}.\delta l$

$$\delta v = 3l^2.\delta l$$

$\therefore$ when $l = 10$ m:

$$\delta v = 3.100.0.01$$

$$\delta v = 3 \text{ m}^3$$

So the volume can vary by 3 m³.

Exercise 11a

1. Find the slope of the curve $y = x^4 + 3x^2 - 7x + 1$ at the point $x = 3$.

2. If the distance in metres $s = t^2 + 3t$, find the velocity and acceleration after 5 seconds if t is measured in seconds.

3. What is the time when the velocity becomes zero for an instant if $s = t^3 - 2t^2$, where t is in seconds and s in metres?

4. If t is the time in seconds and s in metres $= 5t^3 - 3t^2 + 2t + 1$, what is the velocity when the acceleration is zero for an instant?

5. Find the acceleration after 3 m if $v = s^2 + 6s + 7$, where s is in metres, and v in metres per second.

6. If a square is made of sides 15 m within an accuracy of 1 mm, by how much can the area vary from its exact value of 225 m²?

7. A body moves so that its distance from the starting point s in metres $= 12 + 5t + t^3$ after t seconds. What is the distance, velocity and acceleration after 2 seconds?

8. What is the error in the area of a circle and its circumference if its radius is 0.1% greater than its correct radius of 1 m?

11.6

If a curve is plotted from a formula it will probably have turning points such as those shown in the diagram:

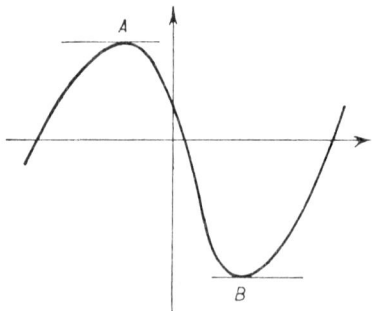

Point A is a maximum and point B is a minimum. At each of these points the tangent is seen to be parallel to the x-axis, i.e. its slope or gradient is zero. The slope is equal to the first derivative, so for a maximum or minimum $\dfrac{dy}{dx} = 0$.

These are two of the turning points or stationary values.

$$\text{Taking } y = x^3 + 4x^2 - 3x + 2$$

$$\frac{dy}{dx} = 3x^2 + 8x - 3$$

for a turning point $0 = (3x - 1)(x + 3)$

i.e. $x = \tfrac{1}{3}, \text{ or } -3$

Thus at $x = \tfrac{1}{3}$ and $x = -3$ there are turning points. We must now find the values of these points.

When $x = \tfrac{1}{3}; \quad y = \tfrac{1}{27} + \tfrac{4}{9} - 1 + 2 = 1\tfrac{13}{27}$

When $x = -3; \quad y = -27 + 36 + 9 + 2 = 20$

11.7

To distinguish between a maximum and minimum we consider how the gradients change as they pass through a turning point. The first

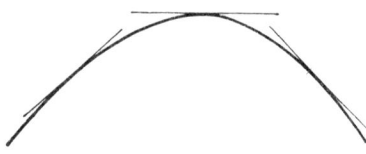

diagram shows a maximum. Before the turning point the slope is positive and after the slope is negative, having passed through zero at the stationary value. Thus the rate of change of gradient is negative. But a rate of change is represented by the derivative

$$\frac{d\left(\frac{dy}{dx}\right)}{dx}, \quad \text{or} \quad \frac{d^2y}{dx^2}$$

The second diagram shows a minimum and it is seen that the slope now changes from negative through zero to positive. Thus the rate of change of slope is positive.

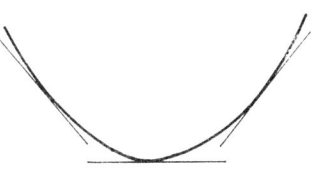

At a turning point $\frac{d^2y}{dx^2}$ is negative for a maximum and positive for a minimum. Taking the example of the previous section:

$$\frac{dy}{dx} = 3x^2 + 8x - 3$$

$$\therefore \frac{d^2y}{dx^2} = 6x + 8$$

When $\qquad x = \frac{1}{3}; \quad \frac{d^2y}{dx^2} = 6.\frac{1}{3} + 8$

which is positive indicating a minimum.

When $\qquad x = -3; \quad \frac{d^2y}{dx^2} = 6.(-3) + 8$

which is negative indicating a maximum.

Thus the curve $y = x^3 + 4x^2 - 3x + 2$ has a maximum of $+20$ at $x = -3$, and a minimum of $1\frac{13}{27}$ at $x = \frac{1}{3}$.

11.8

An alternative method of distinguishing the turning point is to choose x_1 and x_2 on either side of the x_m value for a turning point. The corresponding ordinates y_1 and y_2 are found from the original equation.

If y_1 and y_2 are both less than y_m then it is a maximum. If both are more than y_m it is a minimum.

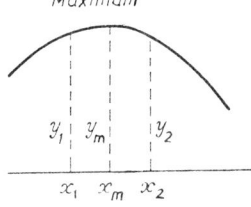

Maximum

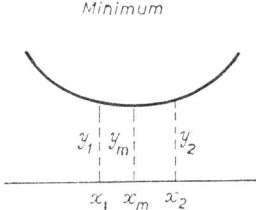

Minimum

E

11.9

In industry it may be important to have a maximum, i.e. for output per man. Also a minimum, i.e. area of sheet required to make a tin of fixed volume.

A can has to have a fixed volume in mm³ of V. It also has to be cylindrical with radius in mm of r and height in mm of h. What relationship between h and r will need the smallest area of tin plate?

$$\text{Area of two ends of can} = 2\pi r^2$$

$$\text{Area of side} = 2\pi rh$$

$$\therefore \text{Area } A = 2\pi(r^2 + rh)$$

but we cannot differentiate until we have removed one of the variables. We can obtain h in terms of r and the constant V

$$V = \pi r^2 h, \quad \text{i.e. } h = \frac{V}{\pi r^2} \tag{1}$$

$$\therefore A = 2\pi r^2 + \frac{2V}{r}$$

$$\frac{dA}{dr} = 4\pi r - \frac{2V}{r^2}$$

For a maximum or minimum

$$0 = 4\pi r - \frac{2V}{r^2}$$

$$\therefore r = \frac{V}{2\pi r^2} = \frac{h}{2},$$

by substitution from (1).

$$\frac{d^2 A}{dr^2} = 4\pi + \frac{4V}{r^3}$$

which is obviously positive giving a minimum area when $2r = h$.

Exercise 11b

1. Find and distinguish the turning point for the curve
$$y = 5x^2 + 6x - 1.$$

2. What stationary value has the curve
$$y = 8x - 5x^2?$$

3. Find the maximum and minimum for the curve
$$y = 2x^3 + 15x^2 + 36x + 1.$$

4. Find the stationary values for the curve $y = x^3 - 3x + 7$.

5. Distinguish and find the values of the turning points of the curve $y = 3x + \dfrac{1}{x}$.

6. Find the turning point for the curve $f(x) = 2x^2 + \dfrac{1}{2x}$.

7. If the curve $y = x^3 + px^2 + qx + 4$ has turning points at $x = 2$ and $x = -1$, find the values of p and q and the nature and values of the turning points.

8. The function $px^3 + qx^2$ has a stationary value of $\frac{1}{2}$ when $x = 1$. Find the values of p and q, the other stationary value, and the nature of both.

9. Find the stationary value of the function $x - \dfrac{4}{x^2}$.

10. The sum of two numbers is 30. Prove that the maximum value of their product is 225.

11. A right-angled triangle has a hypotenuse of 1 m. Find the maximum area, as the other two sides vary.

12. A cone has a fixed volume $V = \frac{1}{3}\pi r^2 h$. Find the value of h in terms of r so that its curved surface area $\pi r(r^2 + h^2)^{1/2}$ is a minimum (i.e. base area excluded).

---

12. Indices, Logarithms, Surds;
the Remainder Theorem

12.1

Powers of numbers are shown by the index of the number; $x.x.x.x$ is written as x^4. We have some basic rules which can be easily demonstrated.

(i) $x^a.x^b = x^{a+b}$. If we write out $x^4.x^3$ fully it is seen to be the same as x^7.

(ii) $\dfrac{x^a}{x^b} = x^a \div x^b = x^{a-b}$. Thus $x^7 \div x^5$ when written out fully cancels down to x^2.

(iii) $(x^a)^b = x^{ab}$ and $(x^b)^a = x^{ab}$. Thus $(x^2)^3$ means
$$(x^2).(x^2).(x^2) = x^{2+2+2} = x^6.$$

(iv) x is so common that it does not have its power put in, but it should be remembered that it is x^1.

(v) $x^0 = 1$. This can be shown by considering $x^4 \div x^4$. Written out

$$\text{this is } \frac{\overset{1}{x} . \overset{1}{x} . \overset{1}{x} . \overset{1}{x}}{\underset{1}{x} . \underset{1}{x} . \underset{1}{x} . \underset{1}{x}} = 1. \text{ By rule (ii) it is } x^{4-4} = x^0.$$

12.2

The powers a and b of the previous section will include roots and reciprocals by applying the following rules.

$\sqrt{x}$ is $x^{1/2}$, $\sqrt[3]{x} = x^{1/3}$, and, in general the nth root of x is written as $x^{1/n}$.

A reciprocal is represented by a negative sign in its power, thus $\frac{1}{x} = x^{-1}$, $\frac{1}{x^2} = x^{-2}$, and in general the reciprocal of the nth power is written as x^{-n}.

With all these fractional and negative powers the rules of section 12.1 apply, e.g.

$$\sqrt[3]{a} \times \sqrt[6]{a} = a^{1/3} . a^{1/6} = a^{(1/3)+(1/6)} = a^{1/2} \text{ or } \sqrt{a}$$

Also $$\sqrt[3]{a^2} = (a^2)^{1/3} = a^{2/3}$$

It does not matter if we find the cube root first and then square, or square first and then take the cube root. Thus $8^{2/3} = \sqrt[3]{64} = 4$, or $8^{2/3} = (\sqrt[3]{8})^2 = (2)^2 = 4$. In general it is best to take the root first in order to deal with smaller numbers.

$$\frac{1}{\sqrt[3]{x^2}} = \frac{1}{(x^2)^{1/3}} = \frac{1}{x^{2/3}} = x^{-2/3}$$

Notice that the reciprocal controls the power's sign and the root its denominator.

Exercise 12a

1. Express $y^3 \div y^{1/2}$ and $(p^{1/2})^{1/3}$ each with a single index and no root sign.

2. Simplify $\left(\dfrac{x^3}{y}\right)^4 \times (-6y^3)^2$.

3. Evaluate $x^{1/2} . x^{-2} . \dfrac{1}{\sqrt[3]{x}}$.

4. Simplify $\sqrt[3]{27x} \div \tfrac{1}{3} . x^2$.

5. Find the values of (a) $(\tfrac{1}{8})^{-1}$, (b) $32^{3/5}$, (c) $(\tfrac{1}{64})^{-2/3}$.

6. Find the value of $27^{1/3} + 9^{-1/3} \cdot 9^{4/3}$.

7. Find the value of n if (a) $8^n = 32$, (b) $9^n = 27$.

8. Simplify $\sqrt[3]{x} \cdot \dfrac{1}{\sqrt{x}} \cdot \dfrac{3}{x^2}$.

12.3

The theory of logarithms follows from the rules of the previous two sections. By expressing numbers as powers of a fixed number we can multiply by adding indices and divide by subtracting indices. The fixed number is called the base and we shall start by considering the common base 10.

$$1000 = 10^3 \text{ is equivalent to writing } \log_{10} 1000 = 3.$$

From our tables we see that $\log_{10} 2 = 0 \cdot 301$ and this is equivalent to $2 = 10^{0 \cdot 301}$ It is the interchange from one of these forms to the other that is the basis of nearly all our work, and helps us to formulate the following rules.

(i) $\log_n a + \log_n b = \log_n (a \cdot b)$.

Let $\qquad \log_n a = x \quad$ and $\quad \log_n b = y$

then $\qquad a = n^x \quad$ and $\qquad b = n^y$

$$a \cdot b = n^x \cdot n^y = n^{x+y}$$

$$\therefore \log_n (a \cdot b) = x + y = \log_n a + \log_n b$$

e.g. $\log_n 6 = \log_n 3 + \log_n 2$.

(ii) $\log_n a - \log_n b = \log_n \left(\dfrac{a}{b}\right)$.

Let $\qquad \log_n a = x \quad$ and $\quad \log_n b = y$

then $\qquad a = n^x$ and $\qquad b = n^y$

$$\frac{a}{b} = \frac{n^x}{n^y} = n^{x-y}$$

$$\therefore \log_n \left(\frac{a}{b}\right) = x - y = \log_n a - \log_n b$$

e.g. $\log_n \frac{3}{2} = \log_n 3 - \log_n 2$.

(iii) $\log_n (a^p) = p \cdot \log_n a$, where p is any power.

Let $\qquad \log_n a = x$

then $\qquad a = n^x \quad \therefore a^p = (n^x)^p = n^{p \cdot x}$

i.e. $\qquad \log_n (a^p) = p \cdot x = p \cdot \log_n a$

e.g. $\log_n \sqrt{a} = \log_n (a^{1/2}) = \frac{1}{2} \log_n a$.

(iv) $\log_n n = 1$.

Let $\qquad\qquad \log_n n = x$

then $\qquad\qquad\qquad n = n^x$

i.e. $x = 1$. The log of a number to its own base is always 1.

(v) $\log_n 1 = 0$.

Let $\qquad\qquad \log_n 1 = x$

then $\qquad\qquad\qquad 1 = n^x$

i.e. $x = 0$. The log of one to any base is zero.

(vi) $\log_n a = \dfrac{\log_m a}{\log_m n}$.

Let $\qquad \log_m a = u$ and $\log_m n = v$

then $\qquad\quad a = m^u$ and $\qquad n = m^v$

i.e. $m = n^{1/v}$.

Eliminating m: $a = (n^{1/v})^u = n^{u/v}$

$$\therefore \log_n a = \frac{u}{v} = \frac{\log_m a}{\log_m n}$$

This rule is useful to convert into logs with a common base, e.g.

$$\log_2 6 = \frac{\log_{10} 6}{\log_{10} 2}$$

Most of these equations are used in both directions. Find the value of x if

$$2\log_{10} x + \log_{10} \tfrac{1}{3} - \log_{10} \tfrac{3}{4} = 2$$
$$\log_{10} x^2 + \log_{10} \tfrac{1}{3} \cdot \tfrac{4}{3} = \log_{10} 100$$
$$\log_{10} x^2 \cdot \tfrac{4}{9} = \log_{10} 100$$
$$\therefore x^2 \cdot 4 = 9 \cdot \overset{25}{\cancel{100}}$$
$$x = \pm 15$$

12.4

When a power is the unknown in an equation it can be solved by taking logarithms.

If $\qquad\qquad\qquad\qquad 2^x = 3$

then $\qquad\qquad\qquad \log_{10} 2^x = \log_{10} 3$

$$x \cdot \log_{10} 2 = \log_{10} 3$$
$$\therefore 0{\cdot}3010 x = 0{\cdot}4771$$
$$x = \frac{0{\cdot}4771}{0{\cdot}3010}$$

If one of these logarithms had involved a bar number it would have had to be made wholly negative. The problem has reduced itself to an ordinary division which can be worked out with the aid of logarithms.

$$x = 1.59$$

No	Log.
0·4771	$\bar{1}$·6786
0·3010	$\bar{1}$·4786
1·585	0·2000

Exercise 12b. (Tables must not be used for 1 to 9.)

1. Show that $2\log\left(\frac{2}{3}\right) + \log\left(\frac{3}{5}\right) - \log\left(\frac{4}{3}\right) = \log\left(\frac{1}{5}\right)$.

2. Find the value of $\dfrac{\log_{10} 32}{\log_{10} 8}$.

3. If $\log_{10} p = 0.7213$, find the value of $\log_{10}\left(\dfrac{1}{p}\right)$, and convert it into a number with the decimal part positive.

4. Given that $\log_{10} a = p + q$ and $\log_{10} b = p - q$, express in terms of p and q the values of (a) $\log_{10} a^3 . b^3$, and (b) $\log_{10}\left(\dfrac{a}{100 . b}\right)$.

5. Find in terms of $\log_{10} 2$ and $\log_{10} 3$ (a) $\log_{10} 6$, (b) $\log_{10}\left(\frac{9}{2}\right)$. Deduce the value of $\log_{10} 3$ given that $\log_{10} 6 = 0.77815$ and $\log_{10} 4.5 = 0.65321$.

6. If $\log_{10}\dfrac{q}{5} - 3\log_{10} p = 2$, obtain q in terms of p.

7. Find the value of $3\log_{10} 2 + \log_{10}\frac{5}{3} - \frac{1}{2}\log_{10}\frac{4}{225}$.

8. Simplify (a) $\log p^5 - \log p^2$, (b) $\log p^5 \div \log p^2$.

9. If $\log_{10} p = -2 + \log_{10} q = \frac{1}{2}\log_{10} 6\frac{1}{4}$, find the values of p and q if both are positive.

10. Find the value of x if $5^x = 26.3$.

11. Solve the equation $3^y = 1.23$.

12. If $(7.21)^p = (126)^3$, find the value of p.

12.5

A surd is a root which will not give an exact answer when the root is extracted. Numbers including surds such as $\sqrt{2}$, $\sqrt[3]{4}$ or $(1 + \sqrt{2})$ are said to be irrational. Other numbers appear to be surds but are rational, for example $\sqrt[3]{8} = 2$; $\sqrt[3]{-8} = -2$; or $\sqrt[4]{16} = \pm 2$. Notice that the odd roots have the same sign as the number but even roots can only be found if the number is positive and give two answers; one positive, the other negative. Some roots will be mixed, for example $\sqrt{8} = \sqrt{4.2} = \sqrt{4}.\sqrt{2} = \pm 2\sqrt{2}$.

12.6

We are able to rationalize the denominators of many numbers. Some of them often occur when trigonometry is used

$$\tan 30° = \frac{1}{\sqrt{3}} = \frac{1}{\sqrt{3}} \cdot \frac{\sqrt{3}}{\sqrt{3}} = \frac{\sqrt{3}}{3}$$

In rationalization we take the root as positive, so that $-\sqrt{3}$ means the negative root only.

When the denominator is the sum of a rational and an irrational part we multiply by its conjugate to give the difference of two squares.

$$\frac{1}{1+\sqrt{2}} = \frac{1}{(1+\sqrt{2})} \cdot \frac{(1-\sqrt{2})}{(1-\sqrt{2})} = \frac{1-\sqrt{2}}{1-2} = \frac{1-\sqrt{2}}{-1} = \sqrt{2}-1$$

This applies to differences and mixed numbers, i.e.

$$\frac{1+\sqrt{2}}{2\sqrt{2}-1} = \frac{(1+\sqrt{2})(2\sqrt{2}+1)}{(2\sqrt{2}-1)(2\sqrt{2}+1)} = \frac{2\sqrt{2}+1+4+\sqrt{2}}{8-1} = \frac{5+3\sqrt{2}}{7}$$

Note that surds such as $\sqrt{2}+\sqrt{3}$ cannot be simplified.

12.7

The remainder theorem states that if any rational integral function $f(x)$ is divided by $(x-a)$, until there is no x term in the remainder, then that remainder $= f(a)$.

$f(a)$ means that the value $x = a$ is substituted in the original function $f(x)$. Suppose on division the quotient is Q and the remainder R, then

$$f(x) = Q(x-a)+R$$

On substituting the value $x = a$ we obtain

$$f(a) = Q(a-a)+R$$
$$= \text{zero}+R \quad \text{i.e.} f(a) = R$$

When $f(a) = 0$ it means that $R = 0$, but if there is no remainder $(x-a)$ must be a factor.

The expression $2x^3+bx^2+cx+2$ has a factor $(x+2)$; and the remainder is 18 when it is divided by $(x-1)$. Find the value of b and c.

By the remainder theorem:

$(x+2)$ factor; substitute $x = -2$:

$$-16+4b-2c+2 = 0 \quad \text{i.e. } 2c = 4b-14$$
$$\therefore c = 2b-7 \tag{1}$$

$(x-1)$ substitute $x = +1$:

$$2+b+c+2 = 18 \quad \text{i.e. } c = -b+14 \tag{2}$$

Eliminating c from (1) and (2):

$$2b - 7 = -b + 14$$
$$3b = 21$$
$$b = 7$$

Substituting in (1), $c = 14 - 7 = 7$, and these values check in equation (2):

$$\therefore b = 7 \quad \text{and} \quad c = 7$$

Exercise 12c. (Consider roots to be positive.)

1. Rationalize the following denominators:

   (a) $\dfrac{1}{\sqrt{2}}$; (b) $\dfrac{2}{\sqrt{3}}$; (c) $\dfrac{6}{\sqrt{3}}$; (d) $\dfrac{\sqrt{2}}{\sqrt{7}}$

2. Rationalize the denominators of:

   (a) $\dfrac{1}{1+\sqrt{3}}$; (b) $\dfrac{1}{\sqrt{3}-1}$; (c) $\dfrac{\sqrt{3}+1}{\sqrt{3}-1}$

3. Prove that $(x+1)$ and $(3x+2)$ are factors of $3x^3 + 2x^2 - 3x - 2$, and find the third factor.

4. The function $x^3 + bx^2 + cx - 6$ has a factor $(x-1)$; and $(x+1)$ leaves a remainder of -24. Find b and c.

5. Rationalize the denominators of:

   (a) $\dfrac{2+3\sqrt{2}}{2-3\sqrt{2}}$; (b) $\dfrac{1}{(2-\sqrt{3})(3+\sqrt{2})}$

6. $x^4 + 3x^3 + ax^2 + bx - 18$ is divisible by $x^2 - 9$ (note this double factor). Find the values of a and b.

7. Simplify $(2 . \sqrt{12} - 3)(2 . \sqrt{3} + 1)$.

8. Show that $(x-1)$ and $(x-2)$ are factors of
   $$x^3 + 3x^2 . (p-2) + (11 - 9p)x + 6(p-1)$$
   for all values of p.

9. Simplify (a) $\sqrt{48}$; (b) $\sqrt{72}$; (c) $\sqrt{300}$.

10. Rationalize the denominator of
    $$\frac{2+4\sqrt{3}}{3\sqrt{2}-5}$$
    and simplify.

11. Simplify (a) $\sqrt{8} + \sqrt{18}$; (b) $\sqrt{27} + \sqrt{45}$; (c) $\sqrt{18} + \sqrt{12}$.

12. Find the values of a and b if $(x+3)$ and $(x-2)$ are factors of $3x^4 + ax^3 - 17x^2 - 13x + b$. Find the other two factors.

13. Integration

13.1

If we differentiate the equation $y = x^2$ we obtain $\dfrac{dy}{dx} = 2x$. Thus the infinitesimal quantity $dy = 2x \cdot dx$. The original quantity y will be the sum of all the quantities dy. This is written as $\int dy$. Also y will be the sum of all the quantities on the right of the second equation, i.e. $\int 2x \cdot dx$. Considering the original equation it is seen that $\int 2x \cdot dx$ gives x^2. Thus the instruction to integrate $2x$ with respect to x, i.e. $\int 2x \cdot dx$ can be accomplished by reversing the process of differentiation.

13.2

In the previous section we were able to integrate back to x^2 because we knew the original equation. In practice we are only given the instruction to integrate $\int 2x \cdot dx$. To do this we use the deduction of the previous section that the answer is a function which when differentiated gives $2x$. As we do not know that original function in practice we must remember that, by section 5.3, added constants when differentiated vanish. So $2x$ could have resulted from $x^2 + a$, $x^2 + b$, $x^2 + \frac{1}{2}$, $x^2 + 3$, $x^2 - 5$, etc. As we are not given any further information we have to put in an undefined added constant which we represent by c.

Thus the integration $\int 2x \cdot dx = x^2 + c$.

13.3

To integrate powers of x we work backwards from the rule of section 2.3.

We shall start with $\int x \, dx$. This has come from one higher power x^2, but if we differentiate this we obtain $2x$, so we must multiply x^2 by $\frac{1}{2}$ to obtain x when differentiated.

$$\therefore \int x \cdot dx = \frac{x^2}{2} + c \quad \text{(our constant of integration)}$$

Added powers will be dealt with separately, so that

$$\int (x + x^2 - x^5) \, dx = \frac{x^2}{2} + \frac{x^3}{3} - \frac{x^6}{6} + c$$

As a check the answer can be differentiated to make sure we get back to the function of x, which we were instructed to integrate.

Exercise 13a. Integrate with respect to x:

1. x^4 2. x^7 3. x^9 4. x^{13}

5. x^{21} 6. $x+x^8$ 7. $x^{10}+x^3$ 8. $x^{12}-x^{11}$

13.4

Fractional and negative powers of x are dealt with in the same way.

$$\int \frac{1}{x^3}.\mathrm{d}x = \int x^{-3}.\mathrm{d}x$$

This has come from one higher power x^{-2}, but if we differentiate this we obtain $-2.x^{-3}$, so we must multiply x^{-2} by $-\frac{1}{2}$ to give $+x^{-3}$ when differentiated, i.e.

$$\int x^{-3}.\mathrm{d}x = -\tfrac{1}{2}.x^{-2}+c = -\frac{1}{2x^2}+c$$

$$\int \sqrt{x}.\mathrm{d}x = \int x^{1/2}.\mathrm{d}x$$

This has come from $x^{3/2}$ which when differentiated gives $\frac{3}{2}.x^{1/2}$. Thus we must multiply $x^{3/2}$ by $\frac{2}{3}$ to differentiate to x^3, i.e.

$$\int x^{1/2}.\mathrm{d}x = \tfrac{2}{3}.x^{3/2}+c$$

Exercise 13b. Integrate with respect to x:

1. x^{-2} 2. x^{-4} 3. $\dfrac{1}{x^7}$ 4. $x^{1/3}$

5. $\sqrt[4]{x}$ 6. $\sqrt[7]{x}$ 7. $\sqrt{x}-\dfrac{1}{x^2}$ 8. $x^6-\dfrac{1}{x^6}$

9. $\dfrac{x^{12}+1}{x^6}$ (compare with Question 8)

13.5

Each operation will be seen to obey the general formula

$$\int x^n.\mathrm{d}x = \frac{x^{n+1}}{n+1}+c$$

The power n can be an integer or fraction, positive or negative with only one exception. n cannot equal -1, for we have never found the result of a differentiation to involve $\frac{1}{x}$. Integrations of $\frac{1}{x}$ are beyond the scope of this book.

$$\int \frac{1}{x.\sqrt{x}}.\mathrm{d}x = \int \frac{1}{x^{3/2}}.\mathrm{d}x = \int x^{-3/2}.\mathrm{d}x = \frac{x^{-1/2}}{-\frac{1}{2}}+c = \frac{-2}{\sqrt{x}}+c$$

13.6

The integration of a constant alone or an added constant will obey the general rule.

$$\int a.dx = \int a.1.dx = \int a.x^0 dx = ax + c$$

Differentiation of this result confirms that it is correct, i.e.

$$\int 3.dx = 3x + c$$

13.7

A multiplied constant when differentiated was shown by section 5.2 to carry through the problem, and so in integration it will carry through.

$$\int ax^3.dx = a.\frac{x^4}{4} + c$$

These multiplied constants can be taken outside the integral thus

$$\int 5x^2.dx = 5 \int x^2.dx = 5.\frac{x^3}{3} + c$$

This will be found much more useful in section 13.9.

Very often the constant will be useful in performing the integration

$$\int 4x^3.dx = x^4 + c$$

This can be used after practice at integration but care must be taken, it is a common mistake to think $2x^3$ or $3x^3$ integrate to x^4. If in doubt, use the previous method

$$\int 4x^3.dx = 4.\frac{x^4}{4} + c = x^4 + c$$

and check by differentiating the answer.

$$\int (3x^2 + ax + b).dx = x^3 + \frac{a.x^2}{2} + bx + c$$

Exercise 13c. Integrate with respect to x:

1. $x^{-5/4}$

2. $x^{-4/5}$

3. $x^{4/5}$

4. $\dfrac{1}{x^2.\sqrt{x}}$

5. 5

6. $21x^2 - 4x + 3$

7. $5x^4 - 6$

8. $\dfrac{1}{\sqrt{x^5}}$

9. $6x^2 + 6 + \dfrac{6}{x^2}$

10. $21x^{21}$

11. $\dfrac{1}{\sqrt[5]{x}}$

12. $17x^{18}$

13.8

There is no formula for the integration of products and so they have to be multiplied out and integrated.

$$\int (1-x)(3+x).dx = \int 3-2x-x^2.dx$$

$$= 3x-x^2-\frac{x^3}{3}+c$$

13.9

There are some types which can be solved by function of a function in reverse. These can also be solved by substitution methods of Chapter 20, but the following method is shorter.

$$\int (4x+3)^2.dx$$

We can spot that this has come from the form $(4x+3)^3$. When we differentiate this by section 9.6 we obtain $3(4x+3)^2.4$. We must therefore introduce a multiplied 12 into the integral

$$\tfrac{1}{12}\int 12.(4x+3)^2.dx$$

The $\tfrac{1}{12}$ is also introduced to keep this new integral equal to the original and has been taken outside because it is a multiplied constant. The integral is now the exact reverse of a differential

$$\therefore \int (4x+3)^2 = \tfrac{1}{12}.[(4x+3)^3]+c = \tfrac{1}{12}.(4x+3)^3+c$$

Taking another example, this time setting it out without explanation

$$\int \frac{5x}{(x^2+1)^3}.dx \qquad (1)$$

$$= \int 5x.(x^2+1)^{-3}.dx \qquad (2)$$

From type

$$f(x) = (x^2+1)^{-2} \qquad (3)$$
$$f'(x) = -2(x^2+1)^{-3}.2x \qquad (4)$$
$$f'(x) = -4x(x^2+1)^{-3} \qquad (5)$$

$$= -\tfrac{5}{4}\int -4.x(x^2+1)^{-3}.dx \qquad (6)$$
$$= -\tfrac{5}{4}(x^2+1)^{-2}+c \qquad (7)$$
$$= \frac{-5}{4(x^2+1)^2}+c \qquad (8)$$

The order of working is shown by the numbers in brackets. Notice that this method could not have been used if there had been no x in the original integral, because *constants* are all that we can introduce into or move outside the integral.

Exercise 13d. Integrate with respect to x:

1. $(3+x)^4$ 2. $3(2-x)^2$ 3. $(5x+4)^4$

4. $x.(x^2+1)^3$ 5. $\dfrac{1}{(2+x)^2}$ 6. $\dfrac{4}{(3-x)^{-5}}$

7. $\dfrac{7x}{(1+x^2)^3}$ 8. $\dfrac{5x^2}{(3+5x^3)^2}$ 9. $\sqrt{3+x}$

10. $x^2\sqrt{1-x^3}$ 11. $\dfrac{1}{\sqrt{1+2x}}$ 12. $(3x^2+2x)(x^3+x^2)^3$

---

14. Addition Formulae; Three-dimensional Problems

14.1

On checking the values of $\sin 50°$, $\sin 30°$, and $\sin 20°$ from the tables it is seen that the first is not the sum of the last two. The correct relationships are given for the angle $(A+B)$ in the following addition formulae which we shall assume:

  (i) $\sin(A+B) = \sin A.\cos B+\cos A.\sin B.$
  (ii) $\sin(A-B) = \sin A.\cos B-\cos A.\sin B.$
  (iii) $\cos(A+B) = \cos A.\cos B-\sin A.\sin B.$
  (iv) $\cos(A-B) = \cos A.\cos B+\sin A.\sin B.$

These can be used as an alternative method of finding the ratios of angles of any magnitude of Chapter 7, using the standard values of $0°$, $90°$, $180°$, $270°$, and $360°$ given there.

$$\begin{aligned}
\sin 210° &= \sin(180°+30°) \\
&= \sin 180°.\cos 30°+\cos 180°.\sin 30° \\
&= 0.\cos 30°+(-1).\sin 30° \\
&= -\sin 30° = -0{\cdot}5000
\end{aligned}$$

This could also have been written $\sin(270° - 60°)$. Also the reciprocal ratios will be solved by this method:

$$\sec 690° = \sec 330° = \frac{1}{\cos 330°} = \frac{1}{\cos(360° - 30°)}$$

$$= \frac{1}{\cos 360°.\cos 30° + \sin 360°.\sin 30°}$$

$$\frac{1}{1.\cos 30°} = \sec 30° = 1·1547$$

The formulae can be used for more theoretical angles, i.e.

$$\sin(-A) = \sin(0 - A) = \sin 0.\cos A - \cos 0.\sin A$$
$$= 0 - (1).\sin A$$
$$= -\sin A$$

or, $\cos(-A) = \cos(0 - A) = \cos 0.\cos A + \sin 0.\sin A$
$$= 1.\cos A = \cos A$$

These could be solved using the quadrant signs, but others are more awkward

$$\sin(270° - A°) = \sin 270°.\cos A° - \cos 270°.\sin A°$$
$$= (-1).\cos A° - 0.\sin A°$$
$$= -\cos A°$$

Exercise 14a. Using the addition formulae find the values of

1. $\sin 170°$ 2. $\sin 240°$ 3. $\sin(90° + 30°)$
4. $\cos 320°$ 5. $\sin 320°$ 6. $\sin 560°$
7. $\cos(270° + 20°)$ 8. $\cos 170°$ 9. $\cos(-170°)$
10. $\sec 120°$ 11. $\csc 220°$ 12. $\sec 310°$

Express as ratios of $A°$:

13. $\sin(90° - A°)$ 14. $\sin(360° + A°)$ 15. $\sin(270° + A°)$
16. $\cos(90° + A°)$ 17. $\cos(360° - A°)$ 18. $\cos(270° + A°)$
19. $\sin(270° - A°)$ 20. $\sin(A° - 180°)$ 21. $\cos(270° - A°)$
22. $\sec(270° + A°)$ 23. $\csc(180° + A°)$ 24. $\csc(-A°)$

14.2

We can use the results of section 14.1 to find $\tan(A + B)$:

$$\tan(A + B) = \frac{\sin(A + B)}{\cos(A + B)} = \frac{\sin A.\cos B + \cos A.\sin B}{\cos A.\cos B - \sin A.\sin B}$$

Dividing both the numerator and denominator by $\cos A \cdot \cos B$ does not change the fraction, i.e.

$$\tan(A+B) = \frac{\dfrac{\sin A \cdot \cancel{\cos B}}{\cos A \cdot \cancel{\cos B}} + \dfrac{\cancel{\cos A} \cdot \sin B}{\cancel{\cos A} \cdot \cos B}}{\dfrac{\cancel{\cos A} \cdot \cancel{\cos B}}{\cancel{\cos A} \cdot \cancel{\cos B}} - \dfrac{\sin A \cdot \sin B}{\cos A \cdot \cos B}}$$

(i) $$\tan(A+B) = \frac{\tan A + \tan B}{1 - \tan A \cdot \tan B}$$

Similarly:

(ii) $$\tan(A-B) = \frac{\tan A - \tan B}{1 + \tan A \cdot \tan B}$$

For example:

$$\tan 265° = \tan(180° + 85°) = \frac{\tan 180° + \tan 85°}{1 - \tan 180° \cdot \tan 85°}$$

$$= \frac{0 + \tan 85°}{1 - 0 \cdot \tan 85°} = \tan 85° = 11 \cdot 43$$

With theoretical angles:

$$\tan(-A) = \tan(0-A) = \frac{\tan 0 - \tan A}{1 + \tan 0 \cdot \tan A} = -\tan A$$

When they include 90° or 270° we cannot use the tangent formula as it involves infinity in the numerator and denominator.

We must therefore split the ratio thus:

$$\tan(90° + A°) = \frac{\sin(90° + A°)}{\cos(90° + A°)} = \frac{\sin 90° \cdot \cos A° + \cos 90° \cdot \sin A°}{\cos 90° \cdot \cos A° - \sin 90° \cdot \sin A°}$$

$$= \frac{1 \cdot \cos A° + 0 \cdot \sin A°}{0 \cdot \cos A° - 1 \cdot \sin A°}$$

$$= -\frac{\cos A°}{\sin A°} = -\cot A°$$

Exercise 14b. Using the addition formulae find:

1. $\tan 110°$ 2. $\tan 200°$ 3. $\tan 280°$

4. $\tan(-50°)$ 5. $\cot 130°$ 6. $\cot(-70°)$

Express as ratios of $A°$:

7. $\tan(180° + A°)$ 8. $\tan(A° - 180°)$ 9. $\cot(180° - A°)$

10. $\cot(-A°)$ 11. $\tan(270° + A°)$ 12. $\cot(90° - A°)$

14.3

When the two parts of the angle are equal we obtain three equations that are very useful.

(i) $\sin (A + A) = \sin A . \cos A + \cos A . \sin A$, i.e.

$$\sin 2A = 2 \sin A . \cos A$$

(ii) $\cos (A + A) = \cos A . \cos A - \sin A . \sin A$, i.e.

$$\cos 2A = \cos^2 A - \sin^2 A$$

but by section 3.4 $\cos^2 A + \sin^2 A = 1$, and this leads to two other forms of the equation which are

$$\cos 2A = 2 \cos^2 A - 1$$
$$\cos 2A = 1 - 2 \sin^2 A$$

(iii) $\tan (A + A) = \dfrac{\tan A + \tan A}{1 - \tan A . \tan A}$

$$\tan 2A = \frac{2 \tan A}{1 - \tan^2 A}$$

To find $\sin 3\theta$ in terms of $\sin \theta$ we first of all apply the results of section 14.1 so that

$$\sin 3\theta = \sin (2\theta + \theta) = \sin 2\theta . \cos \theta + \cos 2\theta . \sin \theta$$

then, using the double angle results:

$$= 2 \sin \theta . \cos^2 \theta . + (1 - 2 \sin^2 \theta) \sin \theta$$
$$= 2 \sin \theta (1 - \sin^2 \theta) + (1 - 2 \sin^2 \theta) \sin \theta$$
$$\sin 3\theta = 3 \sin \theta - 4 \sin^3 \theta$$

We can also use the results for proving relationships which are called trigonometrical identities.

To prove

$$\frac{1 + \sin 2x + \cos 2x}{2 \cos x} = \cos x + \sin x$$

$$\text{L.H.S.} = \frac{1 + 2 \sin x . \cos x + 2 \cos^2 x - 1}{2 \cos x}$$

$$= \frac{2 \cos x (\sin x + \cos x)}{2 \cos x} = \sin x + \cos x = \text{R.H.S.}$$

Exercise 14c

1. Express $\cos 3\theta$ in terms of powers of $\cos \theta$.

2. Express $\cos 4\theta$ in terms of powers of (i) $\cos 2\theta$, (ii) $\cos \theta$.

3. Express $\tan 3\theta$ in terms of powers of $\tan \theta$.

F

Prove the following identities using the various formulae:

4. $\sin(A+B).\cos(A-B)+\cos(A+B).\sin(A-B) = \sin 2A$.

5. $\dfrac{\sin 2A}{1-\cos 2A} = \cot A$.

6. $\cot A - \cot 2A = \operatorname{cosec} 2A$.

7. $\dfrac{1+\cos 2\theta}{1-\cos 2\theta} = \cot^2 \theta$.

8. $(1-\cos 2x+\sin x)\cot x = \sin 2x+\cos x$.

9. $\cos(x+y).\cos(x-y) = \cos^2 y-\sin^2 x$.

10. $\sin(x+y).\sin(x-y) = \sin^2 x-\sin^2 y$.

11. $\cos^4 \theta+\sin^4 \theta = 1-\frac{1}{2}\sin^2 2\theta$.

12. $\cos^4 \theta-\sin^4 \theta = \cos 2\theta$.

14.4

The addition formulae can be used for finding many angles in terms of the standards $30°$, $45°$, $60°$, $90°$, $180°$, $270°$, and $360°$.

For example:

$$\cos 15° = \cos(45°-30°)$$
$$= \cos 45°.\cos 30°+\sin 45°.\sin 30°$$
$$= \frac{1}{\sqrt{2}}.\frac{\sqrt{3}}{2}+\frac{1}{\sqrt{2}}.\frac{1}{2}$$
$$= \frac{1}{2\sqrt{2}}.(\sqrt{3}+1) = \frac{\sqrt{2}(\sqrt{3}+1)}{4}$$

Exercise 14d. Find, without using tables, in surd form:

1. $\sin 15°$	2. $\sec 15°$	3. $\tan 15°$
4. $\sin 75°$	5. $\cos 75°$	6. $\sin 165°$
7. $\tan 135°$	8. $\sin 330°$	9. $\tan 75°$
10. $\tan 105°$	11. $\cos 255°$	12. $\operatorname{cosec} 345°$

14.5

The main part of the three-dimensional problems will involve work in right-angled triangles, but the final calculations may use the sine or cosine rules. The whole figure should be drawn and then each triangle that is used in the calculations should be drawn separately.

The distance from a point to a line or plane always means the perpendicular or shortest distance. The angle formed by two lines is quite definite, but for a line and a plane is more difficult. The correct angle is that between the line and its projection on the plane. The best way to find the projection is to imagine the sun directly over the plane. Then the shadow of the line is its projection.

If AB is the line and AN its projection then $B\widehat{N}A$ is 90°, and the angle between line and plane is $B\widehat{A}N$ where $\cos B\widehat{A}N = \dfrac{AN}{AB}$.

The compass bearing of an inclined point (B) is the bearing of its horizontal projection (N).

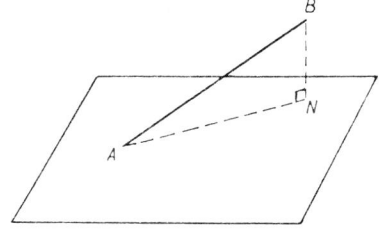

To find the angle between two planes we first of all have to find the

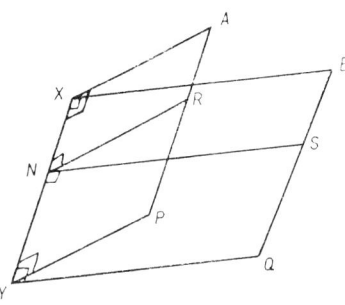

line which is common to both planes, the joining line XY. The angle between the planes is the angle between two lines in the planes which meet each other on XY and are each perpendicular to XY, i.e. $A\widehat{X}B$, $P\widehat{Y}Q$, $B\widehat{N}S$, or an unlimited number of others. Very often the two planes are isosceles (including equilateral) triangles joined at their bases. In this case the most useful angle is $C\widehat{N}D$ between the medians of the two triangles CN and DN. The median to the base of an isosceles triangle bisects the vertical angle and is perpendicular to the base.

When a man is due south of a flagstaff its top is at an elevation of 58°, he walks N. 40° E. (040°) until the elevation of the top is 53°. What is the direction of the flagstaff from the man?

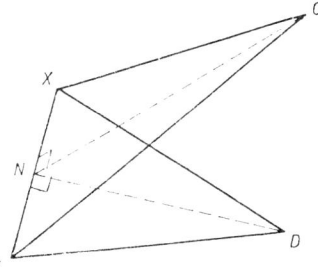

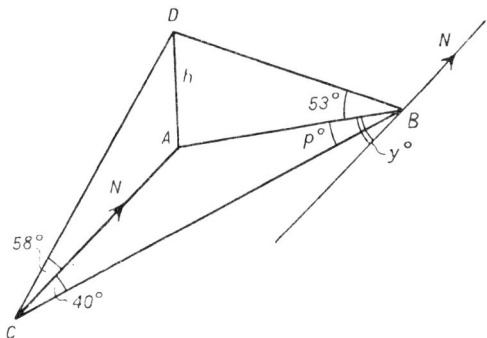

$y°$ is the required angle, and its part $p°$ will enable us to find it. We therefore need as much information as possible about $\triangle ABC$. Two sides can be found in terms of h, the flagstaff height.

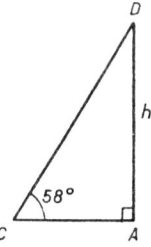 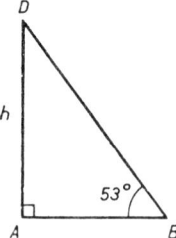

In $\triangle DCA$:

$$CA = h \cdot \cot 58°$$

In $\triangle DAB$:

$$AB = h \cdot \cot 53°$$

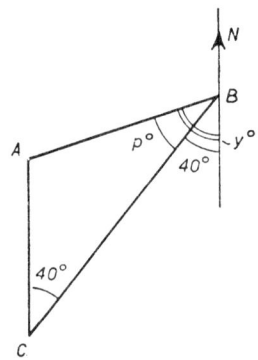

In $\triangle ABC$, by the sine rule:

$$\frac{CA}{\sin p°} = \frac{AB}{\sin 40°}$$

$$\sin p° = \frac{CA \cdot \sin 40°}{AB} = \frac{\cancel{h} \cdot \cot 58° \cdot \sin 40°}{\cancel{h} \cdot \cot 53°}$$

$\bar{1} \cdot 7958$
$\bar{1} \cdot 8081$
$\bar{1} \cdot 6039$
$\bar{1} \cdot 8771$
$\bar{1} \cdot 7268$

$$p° = 32° \, 13'$$

This is the correct value for p since AC is less than AB, so p is less than 40:

$$y° = 40° + 32° \, 13'$$
$$y° = 72° \, 13'$$

The bearing of the flagstaff from the man is S. 72° 13′ W.

Exercise 14e. (Elevations taken from ground level.)

1. Two lamp posts are each 8 m high and an observer notices that

the elevations of the two tops are 10° and 17°. The angle between the two lines from the observer to the base of each lamp post is 54°. Find the distance between the posts.

2. A man observes the elevation of a tree top, which is due south of him, to be 47°. He then walks in a direction S.34°E. past the tree until its elevation is 55°. What is the bearing of the tree from him?

3. A church spire due north of a man, is 50 m high and the elevation of its tip is 50°. The man then walks in a direction N.35°E. until the elevation of the tip of the spire is 17°. How far has the man walked between the two observations?

4. Observer Post O_1 is 8 km south-east of post O_2. An aircraft due north of O_1 is at an elevation of 40°. The elevation of the aircraft from O_2 is 11°. What is the height of the aircraft? (*Hint*: If aircraft is over point B on the ground use the sine rule in $\triangle O_1 O_2 B$.)

5. In a tetrahedron $ABCD$, $BC = BD = 14 \cdot 6$ cm., $AC = AD = 16 \cdot 4$ cm., $AB = 13 \cdot 2$ cm., $CD = 11 \cdot 8$ cm. Prove that, if E is the mid-point of CD, the edge CD is perpendicular to the plane ABE. Calculate the length of the perpendicular from A to face BCD.

6. $ABCD$ is a square base and the point E forms a pyramid such that each face is isosceles, with a vertical angle 35° and equal sides 8 cm. long, calculate (i) the angle between any face and the base, (ii) the angle between AE and the base, (iii) the angle between planes ABE and BCE.

15. Definite Integrals, Areas, Volumes

15.1

The definite integral arises when we consider the area under a curve. If we wish to find the area $ABCD$ it will be the area below the line $y = f(x)$, but above the x-axis, between the lines $x = x_1$ and $x = x_2$. We consider a small element of thickness δx, where as usual this is a small amount compared with x. If it is made small then the element is an oblong of area $y \cdot \delta x$. The total area is the sum of all these smaller

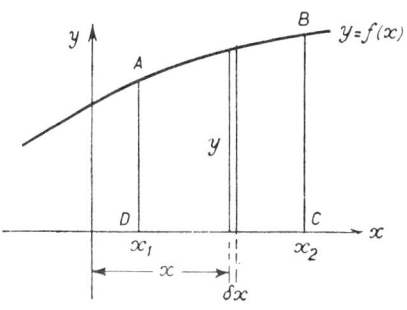

areas, i.e. $\sum y . \delta x$. The exact area is given by the limiting value when $\delta x \to 0$. Thus the area is

$$\lim_{\delta x \to 0} \sum y . \delta x = \int y . dx$$

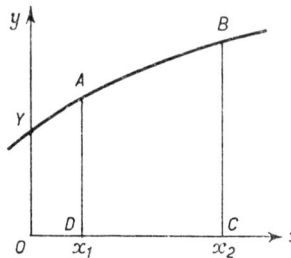

The above formula gives us the area under the curve, above the x-axis, from the y-axis as far as the line $x = x$. Thus if we substitute the value $x = x_2$ we obtain area $OYBC$. Again if we substitute $x = x_1$ we obtain area $OYAD$. The required area is the difference between these two areas, and the instructions to carry this out are written:

$$\text{Area } ABCD = \int_{x_1}^{x_2} y . dx$$

which is a definite integral between limits. A negative answer indicates that the area is under the x-axis. We do not consider areas which are partly above and partly below the axis.

For example we can find the area below the curve $y = -x^2 + 5x - 6$, but above the x-axis. First we must find the limits, which are the points where the curve cuts the x-axis, $y = 0$.

$$0 = x^2 - 5x + 6$$
$$0 = (x - 2)(x - 3)$$

Thus $x_2 = 3$, and $x_1 = 2$.

$$\int_{x_1}^{x_2} y . dx = \int_{2}^{3} -x^2 + 5x - 6 . dx = \left[-\frac{x^3}{3} + \frac{5x^2}{2} - 6x + c \right]_{2}^{3}$$

We now substitute the upper limit, and the lower limit and find the difference in the two values.

$$\text{Area} = \left[-\frac{3^3}{3} + \frac{5 \cdot 3^2}{2} - 6 \cdot 3 + c \right] - \left[-\frac{2^3}{3} + \frac{5 \cdot 2^2}{2} - 6 \cdot 2 + c \right]$$
$$= [-9 + 22\tfrac{1}{2} - 18 + c] - [-2\tfrac{2}{3} + 10 - 12 + c]$$
$$= -4\tfrac{1}{2} + c + 4\tfrac{2}{3} - c$$
$$= \tfrac{1}{6} \text{ sq. units}$$

Notice that the constant cancels out. This will apply to all definite integrals, and so the constant is not put in.

To find the area be-
tween two curves, we first
obtain the limits of in-
tegration by finding where
the curves cross each
other, i.e. we solve them
and obtain the values x_1
and x_2. We find the
areas under $y = f(x)$ and

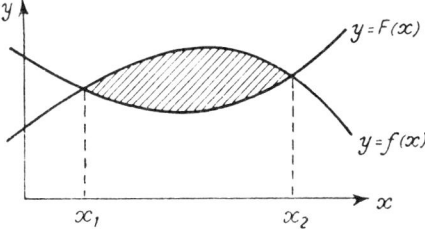

$y = F(x)$ as separate calculations and subtraction gives the shaded area.

Exercise 15a

1. Find the area below the curve $y = -x^2 + 3x - 2$ and above the x-axis.

2. Evaluate $\int_{3}^{6} x^3 - 4 \,.\, dx$.

3. Find the area above the curve $y = x^2 - 4x + 3$ but below the x-axis.

4. Find the area below the curve $y = 1 - x^2$ and above the x-axis.

5. Find the area enclosed between the curves $y = 3x^2$ and $y^2 = 9x$.

6. Find the area between $y = \dfrac{1}{x^2}$, the x-axis, and the limits $x = 2$ and $x = 3$.

7. Evaluate $\int_{0}^{4} x^{3/2} \,.\, dx$.

8. Evaluate $\int_{1}^{2} \dfrac{x^2 + 1}{x^2} \,.\, dx$

9. Show that the area between $y = 2x$ and $y^2 = 4x$ is equal to the area under the line $3y = 2x$ between the same limits. What is the shape of this second area?

10. Find the area between $y = x^3$ and $y = x^4$.

15.2

If the area $ABCD$ turns through one revolution about the portion DC

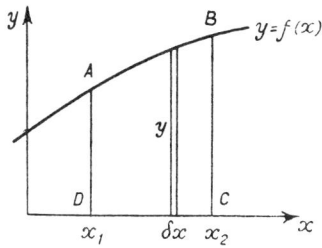

of the x-axis a volume is generated.
To find this volume we first of all
consider the small element of
thickness δx and height y. When
this revolves round the x-axis it
sweeps out a cylindrical disc of
volume $\pi y^2 \,.\, \delta x$ (imagine looking
at a penny edgeways if its radius is
y and the x-axis passes through its

centre). The total volume is the sum of all these small discs, i.e. $\sum \pi y^2 . \mathrm{d}x$. The exact volume is given by the limiting value when $\delta x \to 0$. Thus the volume of revolution of

$$ABCD = \lim_{\delta x \to 0} \sum_{x_1}^{x_2} \pi y^2 . \delta x = \pi \int_{x_1}^{x_2} y^2 . \mathrm{d}x$$

The diagram gives an indication of the volume of the solid of revolution as generated about the x-axis.

The portion AB of the curve can be revolved round the y-axis and the area revolving is now $ABFE$.

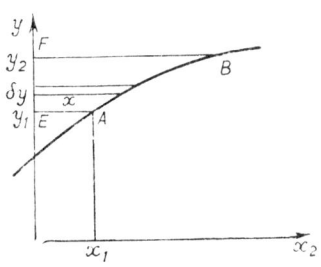

A small element has length x and thickness δy and this forms a cylindrical disc $\pi x^2 . \delta y$. The total volume is $\sum \pi x^2 . \delta y$. The exact volume is given by the limiting value when $\delta y \to 0$. Thus the volume of revolution of

$$ABFE = \pi \int_{y_1}^{y_2} x^2 . \mathrm{d}y$$

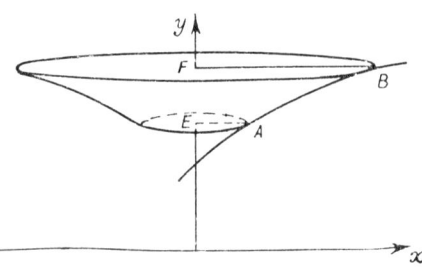

The diagram gives an indication of the volume as generated about the y-axis.

For example let us find the volume of revolution of the parabola $y^2 = 8x$, between the limits $x = 0$ and $x = 2$, (a) about the x-axis, (b) about the y-axis above the x-axis.

(a)   Vol. of revolution $= \pi \int_{x_1}^{x_2} y^2 . \mathrm{d}x = \pi \int_{0}^{2} 8x . \mathrm{d}x$

$$= \pi [4x^2]_0^2 = 4\pi \{[4] - [0]\} = 16\pi \text{ cu. units}$$

(b) When $x = 0$, $y = 0$ and when $x = 2$, $y = 4$ (we take the positive root because of the condition in the question).

$$\text{Vol. of revolution} = \pi \int_{y_1}^{y_2} x^2 . dy = \pi \int_0^4 \frac{y^4}{64} . dy$$

$$= \frac{\pi}{64} \left[\frac{y^5}{5} \right]_0^4 = \frac{\pi}{64.5} \{[4^5] - [0]\} = \frac{\overset{16}{\cancel{4^5}} . \pi}{\cancel{64} . 5}$$

$$= \frac{16\pi}{5} \text{ cu. units}$$

Exercise 15b. (Leave answers in terms of π.)

1. Find the volume of revolution of the parabola $y^2 = 5x$, between the limits $x = 0$ and $x = 1$, (a) about the x-axis, (b) about the y-axis above the x-axis.

2. Find the volume of revolution of $y = 5x$, between the limits $x = 0$ and $x = 6$ about the x-axis. What shape is this volume?

3. The part of the curve $y = x^3$ between the origin and the point (2,8) is rotated about the x-axis. Find the volume of the solid of revolution.

4. By considering the revolution of the line $hy = rx$ about the x-axis between the limits $x = 0$ and $x = h$ (where r and h are constants), find the volume of a cone radius r and height h.

5. Find the volume generated when the area in the first quadrant under the curve $y^2 = 2(3 - x)$, between $x = 0$ and $x = 3$, is rotated through four right angles about the x-axis.

6. The equation of a circle radius a is $y^2 = a^2 - x^2$. By finding the volume of revolution between $x = -a$ and $x = +a$, find the volume of a sphere.

7. Find the volume of revolution when the portion of the curve $y = -x^2 + 5x - 6$ above the x-axis is rotated about that axis.

8. Show that the area enclosed between the straight line $y = x$ and the curve $y = x(2 - x)$ is $\frac{1}{6}$ sq. unit. Find the volume of the solid formed when this area makes a complete revolution about the x-axis, using a method of subtraction similar to that for areas.

16. The Straight Line

16.1

In section 4.3 we found that the straight line could be represented by $y = mx + c$, where m is the slope and c the intercept distance. We then transformed various linear equations to find their m and c. We shall now consider the general equation of the first degree which will represent all such equations. It is $Ax + By + C = 0$. As before we make y the subject:

$$y = -\frac{A}{B}.x - \frac{C}{B}$$

This is a straight line with $m = -\frac{A}{B}$ and $c = -\frac{C}{B}$.

16.2

If the point $P_1(x_1, y_1)$ lies on the straight line $y = mx + c$ we can eliminate c:

$$y = mx + c$$
$$y_1 = mx_1 + c \quad (P_1 \text{ is on the line})$$

i.e. $$y - y_1 = m(x - x_1)$$

16.3

The slope will be found in various ways. If it is parallel to another line it has the same slope. If it is perpendicular to another line its slope is minus the reciprocal of the slope of that line. Both these results were established in section 8.4.

Very often we have to find the slope of a line that passes through two points $P\ (x_1, y_1)$ and $P_2(x_2, y_2)$.

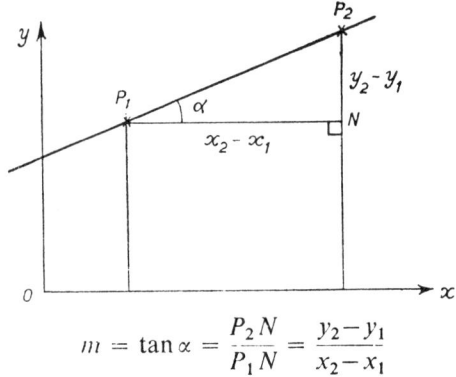

$$m = \tan \alpha = \frac{P_2 N}{P_1 N} = \frac{y_2 - y_1}{x_2 - x_1}$$

This result may be combined with the result of section 16.2 to give an alternative method to section 4.4 for finding the equation of a straight line through two points. Let us again take the points $(1, 2)$ and $(3, -3)$.

$$\text{The slope} = \frac{-3-2}{3-1} = -\frac{5}{2}$$

Then by section 16.2

$$y - 2 = -\frac{5}{2}(x - 1)$$

$$2y - 4 = -5x + 5$$

$$2y = -5x + 9$$

The other point gives the same result:

$$y - (-3) = -\frac{5}{2}(x - 3)$$

$$2y + 6 = -5x + 15$$

$$2y = -5x + 9$$

Exercise 16a. Find the equations of the straight lines which pass through the following pairs of points:

1. $(2, 1)$ and $(3, 3)$.
2. $(2, -1)$ and $(4, 2)$.
3. $(-3, 1)$ and $(-1, -1)$.
4. $(-2, -2)$ and $(3, 3)$.
5. $(-1, -4)$ and $(-2, -3)$.
6. $(2, -3)$ and $(-4, 2)$.
7. $(2, 1)$ and $(8, 1)$.
8. $(3, 2)$ and $(-3, 2)$.

16.4

We shall now consider some special forms of the equation of a straight line.

$x = c$ is a line parallel to the y-axis ($x = 0$)

$y = c$ is a line parallel to the x-axis ($y = 0$)

$y = mx$ goes through the origin (because $c = 0$)

16.5

The straight line cutting off intercept a on the x-axis and intercept b on the y-axis is another form.

Taking any point P on the line, if we can connect its co-ordinates we have the equation of the line.

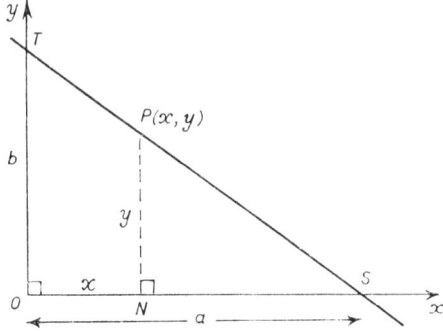

Triangles SNP and SOT are similar $(A.A.A)$

$$\therefore \frac{SN}{SO} = \frac{NP}{OT}$$

i.e.

$$\frac{a-x}{a} = \frac{y}{b}$$

$$\frac{a}{a} - \frac{x}{a} = \frac{y}{b}$$

$$\frac{x}{a} + \frac{y}{b} = 1$$

Alternatively, if the line is $y = mx + c$, then $m = -\dfrac{b}{a}$ and $c = b$, i.e.

$$y = -\frac{b}{a}.x + b \quad \text{or} \quad \frac{x}{a} + \frac{y}{b} = 1$$

Exercise 16b

1. Find the equation of the line through $(2, 1)$ parallel to the line $3y + 2x - 1 = 0$.

2. Find the equation of the line through $(2, 1)$ perpendicular to the line $3y + 2x - 1 = 0$.

3. Find the equation of the line through $(7, -2)$ with gradient $-\frac{1}{2}$.

4. Find the equation of the line through $(2, 1)$: (i) parallel to the x-axis, (ii) parallel to the y-axis, (iii) through the origin.

5. If the intercept on the x-axis is -3 and on the y-axis $+5$, what is the equation of the line?

6. What is the equation of the line with intercept 4 on the x-axis and -2 on the y-axis? What is its slope?

16.6

We shall now find the equation to a straight line in the perpendicular form. The perpendicular from the origin to the line is of length p and makes an angle α with the x-axis. Let P be any point (x, y)

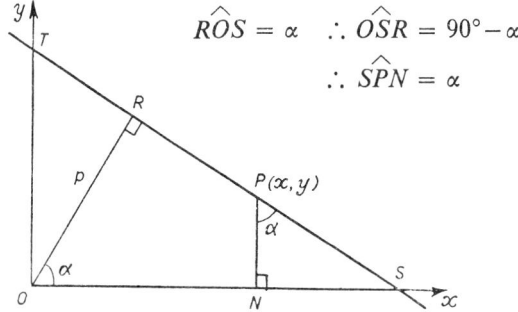

$\widehat{ROS} = \alpha \quad \therefore \widehat{OSR} = 90° - \alpha$

$\therefore \widehat{SPN} = \alpha$

In $\triangle ORS$: $OS = \dfrac{p}{\cos \alpha}$ (1)

In $\triangle PNS$: $NS = y \tan \alpha$ (2)

but $NS = OS - ON = OS - x$ (3)

Eliminating NS from (2) and (3):

$$OS - x = y \tan \alpha \qquad (4)$$

Eliminating OS from (1) and (4):

$$\dfrac{p}{\cos \alpha} = x + y \tan \alpha$$

i.e. $p = x \cos \alpha + y \sin \alpha$, where p is positive since it is an actual positive distance.

For example a line is a perpendicular distance 2 units from the origin and makes an angle of 240° with the x-axis.

The angle is in the third quadrant with basic angle 60°:

$\cos \alpha = -\tfrac{1}{2}$

$\sin \alpha = -\dfrac{\sqrt{3}}{2}$

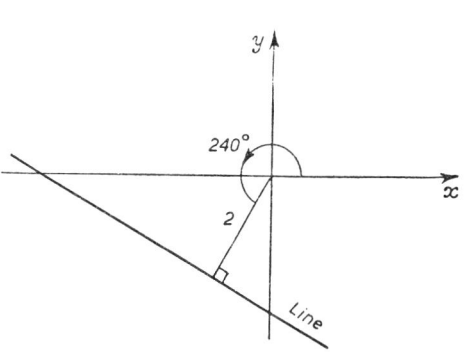

The equation is

$$2 = x.(-\tfrac{1}{2}) + y\left(-\frac{\sqrt{3}}{2}\right) \quad \text{(perpendicular)}$$

$$4 = -x - \sqrt{3}.y$$

$$x + \sqrt{3}y + 4 = 0 \quad \text{(general)}$$

16.7

As the perpendicular form is important we will convert the general first-degree equation

$$Ax + By + C = 0$$

having arranged it to make C positive.

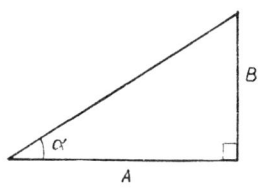

The magnitude of A must represent the magnitude of the numerator of $\cos \alpha$ and B the numerator of $\sin \alpha$. These are represented in the diagram and, by Pythagoras' theorem, the hypotenuse is $+\sqrt{A^2 + B^2}$. We therefore divide each part of the general equation by $+\sqrt{A^2 + B^2}$ to give the correct trigonometrical denominators thus:

$$+C = -Ax - By$$

$$+\frac{C}{\sqrt{A^2 + B^2}} = -\frac{A}{\sqrt{A^2 + B^2}}.x - \frac{B}{\sqrt{A^2 + B^2}}.y$$

which compares with $p = x\cos\alpha + y\sin\alpha$.

Convert the line $12x - 5y + 26 = 0$ to the perpendicular form. The constant term becomes p and must therefore be kept positive, i.e. $26 = -12x + 5y$:

Magnitude $A = +12, \quad B = +5$

then $\sqrt{A^2 + B^2} = 13$

$$\frac{26}{13} = -\frac{12}{13}.x + \frac{5}{13}.y$$

i.e. $$2 = -\frac{12}{13}.x + \frac{5}{13}.y$$

which is in the perpendicular form such that

$$\cos \alpha = -\frac{12}{13} \quad \text{and} \quad \sin \alpha = \frac{5}{13}$$

putting α in the second quadrant.

Exercise 16c. (Leave as surds if they occur.) Find the equations of lines with the following perpendicular distances from the origin and angles with the positive x-axis in (i) the perpendicular form, (ii) the general form.

1. 2 units, 30°.
2. 4 units, 135°.
3. 3 units, 180°.
4. 1 unit, 300°.

Convert to the perpendicular form:

5. $3x + 4y - 15 = 0$.
6. $x + y - 8 = 0$.
7. $-\sqrt{3}x + y - 7 = 0$.
8. $8x - 15y + 34 = 0$.

16.8

To find the perpendicular distance from the point $P_1(x_1, y_1)$ on to a line, we start with the perpendicular form $p = x\cos\alpha + y\sin\alpha$.

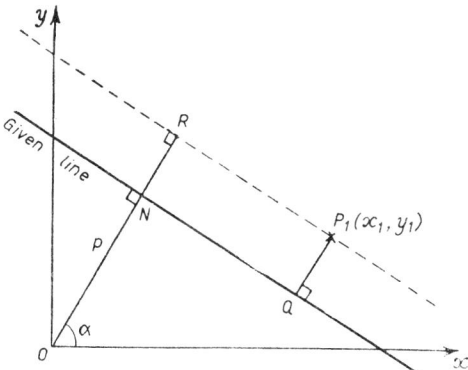

A line is constructed through P_1 parallel to the given line so that the required distance is $P_1 Q$. The perpendicular from the origin meets the given line at N and the constructed line at R so that $RN = P_1 Q$.

The equation of the given line through N and Q is

$$p = x\cos\alpha + y\sin\alpha$$

Let $OR = p'$, then the equation of the constructed line is

$$p' = x\cos\alpha + y\sin\alpha$$

But $P_1(x_1, y_1)$ lies on this line and so must satisfy the equation, thus
$p' = x_1 \cos \alpha + y_1 \sin \alpha$.

We require

$$P_1 Q = RN = p' - p = x_1 \cos \alpha + y_1 \sin \alpha - p$$

$$P_1 Q = x_1 \cos \alpha + y_1 \sin \alpha - p$$

Since p is positive, if the answer is positive P_1 and the origin are on opposite sides of the line, as in the diagram. If it is negative, they are on the same side.

For example find the perpendicular distance from $(4, \sqrt{3})$ on to the line

$$2 = -\tfrac{1}{2}.x - \frac{\sqrt{3}}{2}.y$$

The distance is

$$x_1 \cos \alpha + y_1 \sin \alpha - p = 4.(-\tfrac{1}{2}) + \sqrt{3}.\left(-\frac{\sqrt{3}}{2}\right) - 2$$

$$= -2 - \frac{3}{2} - 2$$

$$= -5\tfrac{1}{2} \text{ units}$$

i.e. $5\tfrac{1}{2}$ units, since $(4, \sqrt{3})$ must be the on same side as the origin.

16.9

To find the perpendicular distance from $P_1(x_1, y_1)$ on to the line $Ax + By + C = 0$, where the constant C is arranged to be positive.

By the previous section

$$P_1 Q = x_1 \cos \alpha + y_1 \sin \alpha - p$$

By section 16.7

$$\cos \alpha = \frac{-A}{\sqrt{A^2 + B^2}}, \sin \alpha = \frac{-B}{\sqrt{A^2 + B^2}}, p = \frac{C}{\sqrt{A^2 + B^2}}$$

thus
$$P_1 Q = \frac{-Ax_1}{\sqrt{A^2 + B^2}} + \frac{-By_1}{\sqrt{A^2 + B^2}} - \frac{C}{\sqrt{A^2 + B^2}}$$

$$P_1 Q = \frac{-(Ax_1 + By_1 + C)}{\sqrt{A^2 + B^2}}$$

Again, if the answer is positive, the point P_1 is on the opposite side to the origin.

For example to find the distance from $(5, 5)$ to the line

$$7x + 24y - 20 = 0$$

Arranging with C positive $-7x - 24y + 20 = 0$

$$\sqrt{A^2 + B^2} = 25$$

$$\therefore \frac{-7x - 24y + 20}{25} = 0$$

Then the perpendicular distance

$$P_1 Q = -\frac{(-7.5 - 24.5 + 20)}{25}$$

$$= \frac{5}{25}(7 + 24 - 4) = \frac{27}{5} = 5\tfrac{2}{5} \text{ units}$$

Exercise 16d. (Leave as surds if they occur.) Find the perpendicular distance to the following lines from the points given:

1. A line 3 units from the origin at an angle $30°$; point $(3, 1)$.
2. A line 2 units from the origin at an angle $330°$; point $(-3, 2)$.
3. A line 5 units from the origin at angle $225°$; point $(0, 3)$.
4. $3x + 4y - 10 = 0$; point $(5, 5)$.
5. $x + y - \sqrt{2} = 0$; point $(-2, 2)$.
6. $-12x - 5y + 13 = 0$; point $(-26, 13)$.
7. $x + \sqrt{3}y + 6 = 0$; point $(4, -2)$.
8. $-15x + 8y + 17 = 0$; point $(51, 34)$.

---

17. Progressions; the Binomial Expansion

17.1

An arithmetical progression (A.P.) is built up by adding a constant amount to the previous term. One of the simplest is 1, 2, 3, 4, 5, ...

We need only two facts to build up the series, we let the first term $T_1 = a$. The amount added (which of course includes subtraction) is d, the common difference. The series is then:

$$T_1, \quad T_2, \quad T_3, \quad T_4, \quad T_5, \quad \dots$$

$$a, \quad a+d, \quad a+2d, \quad a+3d, \quad a+4d, \quad \dots$$

G

A pattern emerges which enables us to write down the general term; we shall call this the nth term, T_n, and it will represent all the terms in the series. We see each term has a and the number of common differences is one less than the number of the term. Thus $T_5 = a + 4d$ and if we worked on $T_{17} = a + 16d$. By the same reasoning $T_n = a + (n-1)d$.

Notice the importance of the general term, it will give all the terms by putting $n = 1, 2, 3, 4, 5$, etc.

The fifth term of an A.P. is 27 and the ninth term is 39. What is the value of the seventh term?

$$T_9 = a + 8d = 39$$

$$T_5 = a + 4d = 27$$

$$\text{Subtract } 4d = 12, \quad d = 3$$

From the first equation $a + 24 = 39$, i.e. $a = 15$. These values check in the second equation.

$$T_7 = a + 6d = 15 + 18 = 33$$

or
$$T_7 = T_5 + 2d = 27 + 6 = 33$$

As a final check the series is:

$$T_1, \quad T_2, \quad T_3, \quad T_4, \quad T_5, \quad T_6, \quad T_7, \quad T_8, \quad T_9, \quad T_{10}, \ldots$$

$$15, \quad 18, \quad 21, \quad 24, \quad 27, \quad 30, \quad 33, \quad 36, \quad 39, \quad 42,$$

Also
$$T_n = 15 + (n-1).3 = 3(n+4)$$

The next step is to find the sum of the first n terms of an A.P.

$$S_n = T_1 + \quad T_2 \quad + \quad T_3 \quad + \ldots \quad T_{(n-2)} \quad + \quad T_{(n-1)} \quad +$$
$$S_n = a + (a+d) + (a+2d) + \ldots \{a + (n-3).d\} + \{a + (n-2).d\} +$$

$$+ \quad T_n$$
$$+ \{a + (n-1).d\}$$

We now notice that $T_1 + T_n = T_2 + T_{(n-1)}$, etc. Any of these pairs equals $2a + (n-1)d$.

If n is even we shall have $\dfrac{n}{2}$ pairs, so $S_n = \dfrac{n}{2}\{2a + (n-1)d\}$.

If n is odd we shall have $\dfrac{n-1}{2}$ pairs, plus the middle term.

$$T_{(n+1)/2} = a + \left(\frac{n+1}{2} - 1\right).d = a + \left(\frac{n+1-2}{2}\right).d = \frac{2a + (n-1).d}{2}$$

Then
$$S_n = \frac{n-1}{2}\{2a+(n-1).d\}+\frac{2a+(n-1).d}{2}$$

$$= \frac{(n-1+1)}{2}\{2a+(n-1).d\} = \frac{n}{2}\{2a+(n-1).d\}$$

Thus the sum of n terms

$$S_n = \frac{n}{2}\{2a+(n-1).d\}$$

Also, $S_n = \frac{n}{2}\{T_1+T_n\}$, or any pair of terms that are equal to their sum.

Find the sum of all the numbers between 1 and 100 divisible by 3.

$$T_1 = 3, \quad T_2 = 6, \quad T_3 = 9 \ldots T_n = 99$$

$$\therefore T_1 = a = 3$$

$$T_2 = a+d = 6 \quad \therefore d = 3$$

$$T_n = a+(n-1).d = 99$$

i.e. $$3+(n-1)3 = 99$$

$$n = 33$$

$$S_n = \frac{33}{2}\{3+99\} = 33 \times 51 = 1683$$

Exercise 17a

1. The second and fifth terms of an A.P. are 7 and 19. What is the tenth term and the sum of the first ten terms?
2. Find the sum of the terms between 50 and 101 divisible by 4.
3. The first term of an A.P. is 26 and the fourth is 11. Find the value of the eleventh term.
4. The nth term of an A.P. has a value $(2n-1)$. Give the first three terms of the series and the sum of the first twelve terms.
5. The sum of an A.P. is 144, the first term is 3 and final term 15. What is the common difference?
6. The fifteenth term of an A.P. is twice the seventh term. Prove that the fourteenth is three times the fourth term.
7. Find p, q, and r such that $15, p, q, r, -3$ form an A.P.
8. The sum of the first and last terms of an A.P. is 42. The sum of all the terms is 420. The second term is 4. What is the common difference?

17.2

A geometrical progression (G.P.) is built up by multiplying the previous term by a constant amount. One of the simplest is 1, 2, 4, 8, 16...

The two facts needed for a G.P. are the first term $T_1 = a$ and the multiplied amount (which includes division), which is r, the common ratio.

The series is then:

$$T_1, \quad T_2, \quad T_3, \quad T_4, \quad T_5, \quad \ldots$$
$$a, \quad ar, \quad ar^2, \quad ar^3, \quad ar^4, \quad \ldots$$

The general or nth term T_n must be ar^{n-1}; for example $T_{17} = ar^{16}$. We can build up the series from the general term by substituting $n = 1, 2, 3, 4, 5$, etc.

The fifth term of a G.P. is $20\frac{1}{4}$ and the third is 9. What is the series, giving the first five terms and the general term?

$$T_5 = ar^4 = \frac{81}{4}$$

$$T_3 = ar^2 = 9$$

Dividing $\qquad r^2 = \frac{81}{4} \times \frac{1}{9} = \frac{9}{4} \quad \therefore r = \pm\frac{3}{2}$

From the first equation

$$ar^4 = a.\frac{81}{16} = \frac{81}{4} \quad \text{i.e. } a = 4$$

These values check in the second equation and lead to two series:

$$T_1, \ T_2, \quad T_3, \ T_4, \qquad T_5 \quad \ldots \ T_n$$

$$4, \ 6, \quad 9, \ 13\frac{1}{2}, \qquad 20\frac{1}{4}, \ \ldots \ 4.\left(\frac{3}{2}\right)^{n-1}$$

or $\qquad 4, \ -6, \ 9, \ -13\frac{1}{2}, \ 20\frac{1}{4}, \ \ldots \ 4\left(-\frac{3}{2}\right)^{n-1}$

To find the sum of the first n terms:

$$S_n = a + ar + ar^2 + ar^3 \ldots + ar^{n-2} + ar^{n-1}$$

$$r.S_n = \qquad ar + ar^2 + ar^3 + ar^4 \ldots \quad + ar^{n-1} + ar^n$$

Subtract:

$$(1-r)S_n = a \qquad\qquad\qquad\qquad\qquad -ar^n$$

$$(1-r)S_n = a(1-r^n)$$

$$S_n = \frac{a(1-r^n)}{(1-r)}$$

Find how many terms of the series 4, 6, 9 ... are needed to make the sum just greater than 1000. Let us find n for the sum to be exactly 1000

$$1000 = \frac{4\{1-(\frac{3}{2})^n\}}{-\frac{1}{2}}$$

$$-125 = 1-\left(\frac{3}{2}\right)^n$$

$$\left(\frac{3}{2}\right)^n = 126$$

| 0·3222
|‾1·2457
|1·0765

$$n.\log 1·5 = \log 126$$

$$n = \frac{\log 126}{\log 1·5} = \frac{2·1004}{0·1761} = 11·93$$

Since n must be a whole number it must be 12 for the sum to be just over a 1000.

Exercise 17b

1. The fifth and second terms of a G.P. are $5\frac{1}{16}$ and $1\frac{1}{2}$. What is the fourth term and the sum of five terms?

2. Find the sum of ten terms of the G.P. $1, 1·2, 1·44 ...$

3. Find the number of terms which will make the G.P. $1, \frac{1}{2}, \frac{1}{4} ...$ just greater than $1·995$.

4. The first two terms of a series are $1, -2$. Find (i) the sum of ten terms if it is an A.P., (ii) the seventh term if it is a G.P.

5. The first and second terms of a G.P. are 192 and -24. How many terms are there if the last one is $\frac{-3}{512}$?

6. The nth term of a G.P. is 2^{n-2}. What is (a) the first term, (b) the second term, (c) the sum of 10 terms?

7. The perimeter of a triangle is 19 cm. The shortest side is 4 cm. What are the lengths of the other two sides if the sides are in (i) A.P., (ii) G.P.

8. The second, fifth and fourteenth terms of an A.P. are in G.P. What is the common ratio of the G.P.? $(d \neq 0)$.

17.3

The results of the binomial theorem of Chapter 1 can be applied to fractional or negative indices. The main differences are that the series are continuous and only true for x values between -1 and $+1$.

To find the first six terms of the series $\dfrac{1}{\sqrt{1+x}}$:

$$\frac{1}{\sqrt{1+x}} = (1+x)^{-1/2} = 1 + (-\tfrac{1}{2}).x + \frac{\left(-\tfrac{1}{2}\right)\left(-\tfrac{3}{2}\right)}{1.2}.x^2 +$$

$$+ \frac{\left(-\tfrac{1}{2}\right)\left(-\tfrac{3}{2}\right)\left(-\tfrac{5}{2}\right)}{1.2.3}.x^3 +$$

$$+ \frac{\left(-\tfrac{1}{2}\right)\left(-\tfrac{3}{2}\right)\left(-\tfrac{5}{2}\right)\left(-\tfrac{7}{2}\right)}{1.2.3.4}.x^4 +$$

$$+ \frac{\left(-\tfrac{1}{2}\right)\left(-\tfrac{3}{2}\right)\left(-\tfrac{5}{2}\right)\left(-\tfrac{7}{2}\right)\left(-\tfrac{9}{2}\right)}{1.2.3.4.5}.x^5$$

$$= 1 - \frac{x}{2} + \frac{3x^2}{8} - \frac{5x^3}{16} + \frac{35x^4}{128} - \frac{63x^5}{256}$$

Exercise 17c

1. Find the first five terms, in ascending order, of the expansion of $\dfrac{1}{(1+x)^2}$.

2. Find the first six terms of the expansion of $\sqrt{1+x}$.

3. Find the first five terms of the expansion of $\dfrac{1}{\sqrt[3]{1+x}}$.

4. Find the first five terms of the expansion of $\dfrac{1}{(1-x)^4}$.

5. Find the first four terms of the expansion of $\dfrac{1}{\sqrt{a+x}}$.

   (Assume $|x| < |a|$ where $|x|$ means the value of x with a positive sign.)

6. Find the first eight terms of the expansion of $\dfrac{1}{(1-x)}$.

7. Find $\sqrt{1 \cdot 02}$ correct to five decimal places using the binomial theorem.

8. Find $\dfrac{1}{\sqrt{0\cdot99}}$ correct to five decimal places using the binomial theorem.

---

18. Trigonometrical, Quotient, and Implicit Functions

18.1

We shall assume two results:

$$y = \sin x \qquad\qquad\qquad y = \cos x$$

$$\frac{dy}{dx} = \cos x \qquad\qquad\qquad \frac{dy}{dx} = -\sin x$$

We must also note that x is in radians unless stated otherwise.

If $y = b\sin(ax)$, where a is a constant the right-hand side is a function of a function

$$y = b\sin u \qquad\qquad\qquad u = ax$$

$$\frac{dy}{du} = b\cos u \qquad\qquad\qquad \frac{du}{dx} = a$$

$$\therefore \frac{dy}{dx} = b(\cos u)\,a = ab\cos ax$$

This can be done in one step.

$$y = \sin x^\circ = \sin\left(\frac{\pi x}{180}\right)$$

$$\therefore \frac{dy}{dx} = \frac{\pi}{180}\cos\left(\frac{\pi x}{180}\right) = \frac{\pi}{180}\cos x^\circ$$

Exercise 18a. Differentiate with respect to x:

1. $\cos ax$ 2. $\cos x^\circ$ 3. $\sin(2x)$

4. $2\cos(3x)$ 5. $\sin(x^2)$ 6. $\cos(ax+b)$

7. $\sin(ax^2+bx)$ 8. $\sin^2 x$ (i.e. $(\sin x)^2$) 9. $\cos^3 x$

18.2

To find the differential of $\tan x$ we transform it to

$$\frac{\sin x}{\cos x} = \sin x.(\cos x)^{-1}$$

$$y = \sin x.(\cos x)^{-1}$$

then by the product rule

$$\frac{dy}{dx} = \sin x.(-1).(\cos x)^{-2}.(-\sin x) + (\cos x)^{-1}.\cos x$$

$$= \frac{\sin^2 x}{\cos^2 x} + 1 = \tan^2 x + 1 = \sec^2 x$$

Again, if $y = \sec x$ we have

$$y = (\cos x)^{-1}$$

$$\therefore \frac{dy}{dx} = -1.(\cos x)^{-2}.(-\sin x)$$

$$= \frac{\sin x}{\cos x}.\frac{1}{\cos x} = \tan x.\sec x$$

Using this result, if $y = \sec ax$, then

$$\frac{dy}{dx} = a.\tan ax.\sec ax$$

Exercise 18b. Differentiate with respect to x:

1. Assuming the differentials already given, (a) $\operatorname{cosec} x$, (b) $\cot x$.
2. $\tan(ax)$ 3. $\cot(bx^2)$ 4. $x.\sin 2x$
5. $\sec x + \tan x$ 6. (a) $\tan^2 x$, (b) $\sec^2 x$
7. $(1 + \sec x)^3$ 8. $\sec^2 x.\operatorname{cosec}^2 x$ 9. $\sec^2 x + \operatorname{cosec}^2 x$
10. Find the maximum and minimum values of $y = \sin x$. $(0 < x < 2\pi.)$

18.3

To differentiate a quotient $y = \dfrac{u}{v}$ we can convert it into a product and apply the product rule:

$$y = u.v^{-1}$$

where u and v are functions of x.

$$\frac{dy}{dx} = u.(-1).v^{-2}.\frac{dv}{dx} + v^{-1}.\frac{du}{dx}$$

$$= -\frac{u}{v^2}.\frac{dv}{dx} + \frac{1}{v}.\frac{du}{dx}$$

This is usually written

$$\frac{dy}{dx} = \frac{v \cdot \frac{du}{dx} - u \cdot \frac{dv}{dx}}{v^2}$$

This formula can be applied as a standard but care must be taken to get the numerator in the right order. If $y = \frac{x}{\sin x}$ it is usual to start with the denominator of the derivative ($\sin^2 x$) and then start the numerator with the original denominator ($\sin x$), thus

$$\frac{dy}{dx} = \frac{\sin x \cdot 1 - x \cdot \cos x}{\sin^2 x} = \frac{\sin x - x \cdot \cos x}{\sin^2 x}$$

Exercise 18c. Differentiate with respect to x:

1. $\dfrac{1+x}{1-x}$ 2. $\dfrac{(2-3x)}{(1+x^2)}$ 3. $\dfrac{(1+x)^3}{(1+x^3)}$

4. $\dfrac{(1+x)^2}{(1-x)^2}$ 5. $\dfrac{x^2}{\tan x}$ 6. $\dfrac{\sin x}{x^3}$

7. $\dfrac{\sin x}{\cos x}$ (check for $\tan x$) 8. $\dfrac{\tan x}{\sec x}$ (i.e. $\sin x$)

9. $\dfrac{5x+7}{3x^2+1}$ 10. $\dfrac{\sqrt{x^3}}{(1-x)}$ 11. $\dfrac{\sqrt{1+x}}{\sqrt{1-x}}$

12. Find the maximum and minimum values of $y = \dfrac{3+x^2}{1+x}$.

18.4

In section 9.4 we have seen that we can deal with an expression without making y the subject. Thus $y^2 = x^3$ is a function in the implicit form. It becomes explicit as $y = x^{3/2}$. But there are some functions which can only be implicit such as $y + \tan y = x^3$.

We have seen that the first can be differentiated by the function of a function rule: $y^2 = x^3$.

$$\frac{d(y^2)}{dx} = 3x^2$$

$$\frac{d(y^2)}{dy} \cdot \frac{dy}{dx} = 3x^2 \quad \text{i.e.} \quad 2y \cdot \frac{dy}{dx} = 3x^2, \quad \frac{dy}{dx} = \frac{3x^2}{2y}$$

(If y is eliminated we obtain the same answer as the derivative of $y = x^{3/2}$).

In the same way we have
$$y + \tan y = x^3$$
$$\frac{d(y + \tan y)}{dy} \cdot \frac{dy}{dx} = 3x^2$$

i.e. $(1 + \sec^2 y) \cdot \dfrac{dy}{dx} = 3x^2; \quad \dfrac{dy}{dx} = \dfrac{3x^2}{(1 + \sec^2 y)}$

Thus it is seen that to differentiate with respect to x when the term is in y we have to differentiate with respect to y and multiply by $\dfrac{dy}{dx}$.

Many implicit functions have products which contain x and y and as an example we shall take
$$x^2 + x^3 . y^2 + y = 0$$

The derivative of x^2 is $2x$.

The derivative of $x^3 . y^2$ by the product rule is
$$x^3 . 2y . \frac{dy}{dx} + y^2 . 3x^2$$

The derivative of y is $\dfrac{dy}{dx}$.

so, if $x^2 + x^3 . y^2 + y = 0$
$$2x + 2x^3 . y . \frac{dy}{dx} + 3x^2 y^2 + \frac{dy}{dx} = 0$$

i.e. $(2x^3 . y + 1) \dfrac{dy}{dx} = -x(2 + 3xy^2)$
$$\frac{dy}{dx} = \frac{-x(2 + 3xy^2)}{(1 + 2x^3 . y)}$$

Exercise 18d. Differentiate with respect to x:

1. $y^2 + \sin y = \cos x$.
2. $y . \sin y = x$.
3. $y^2 = x^2 - 2xy$.
4. $0 = (3 - x)(4 + y)$.
5. $y = x . \sin y + x^2$.
6. $\tan y = x^2 y^3$.
7. $\cos y = \dfrac{y^4}{x^2}$.

8. Calculate the values of $\dfrac{dy}{dx}$, when $x = 1$, for the equation
$$4y^2 - 7xy + x^2 = 3.$$

19. Simultaneous and Quadratic Equations

19.1

When we have three unknowns in an equation, such as $x+y+z = 0$, we shall need two other equations in order to solve for x, y, and z.

For example

$$x+y+z = 0 \tag{1}$$

$$2x-y+3z = 11 \tag{2}$$

$$3x-2y-5z = -1 \tag{3}$$

We now take the equations in two pairs and eliminate the same letter from both. Let us eliminate y:

Add (1) and (2):

$$3x+4z = 11 \tag{4}$$

(1) $\times$ 2: $2x+2y+2z = 0$
(3): $3x-2y-5z = -1$

Adding the pair:

$$5x-3z = -1 \tag{5}$$

Equations (4) and (5) are now two simultaneous equations in two unknowns that we have met before. Let us eliminate z

(4) $\times$ 3: $9x+12z = 33$
(5) $\times$ 4: $20x-12z = -4$

Adding: $29x = 29$

$$x = 1$$

Substituting x in (4):

$$4z = 11-3 = 8$$

$$z = 2 \quad \text{(Check in (5))}$$

Substituting x and z in (1):

$$y = -1-2 = -3$$

$$y = -3 \quad \text{(Check in (2) and (3))}$$

The solution is $x = 1$, $y = -3$, $z = 2$.

Example 19a. Solve the following simultaneous linear equations:

1. $2x + y + z = 7$
   $x - 4y + 2z = -1$
   $x + 3y + 3z = 16$

2. $p + q + r = 3$
   $3p - 3q + 2r = 13$
   $2p + q - 3r = -3$

3. $x + y + z = 5$
   $3x + 2y + 2z = 13$
   $2x - 5y = -14$

4. $2p - 3q + 4r = 3$
   $4p - 5q - r = -2$
   $p + q + r = 3$

5. $l + m - 4n = 0$
   $2l + 3m + n = -3$
   $5l + 6m + 4n = -3$

6. $x + y + z = 0$
   $2x + 3y + 4z = 9$
   $2(x + z) = 3(y - 5)$

19.2

If we have a linear and a second-degree equation in two unknowns we start with the linear equation making one of the unknowns the subject,

$$2x - \frac{(3 + y)}{5} = 1 \tag{1}$$

$$x^2 + 2xy + y^2 = 9 \tag{2}$$

Equation (1) is linear although it is rather disguised, i.e.

$$10x - 3 - y = 5$$
$$10x - 8 = y \tag{3}$$

This value of y is then substituted in (2)

$$x^2 + 2x(10x - 8) + (10x - 8)^2 = 9$$
$$x^2 + 20x^2 - 16x + 100x^2 - 160x + 64 = 9$$
$$121x^2 - 176x + 55 = 0$$

Divide by 11:

$$11x^2 - 16x + 5 = 0$$
$$(11x - 5)(x - 1) = 0 \quad \therefore x = 1 \text{ or } \tfrac{5}{11}$$

We now substitute these in equations (3):

When $\qquad x = 1, \quad y = 10 - 8 = 2$

When $\qquad x = \tfrac{5}{11}, \quad y = \tfrac{50}{11} - 8 = -3\tfrac{5}{11}$

These values are checked in (2).

When $x = 1$, $y = 2$, and when $x = \tfrac{5}{11}$, $y = -3\tfrac{5}{11}$.

Exercise 19b. Solve the following simultaneous equations:

1. $3x+y = 4$
   $x^2 - xy + 3y^2 = 3$

2. $5x + 2y = 8$
   $2x^2 + 3xy = 2$

3. $8y + x = 2$
   $x^2 + xy - 4y^2 = 2$

4. $5x = 2(y-1)$
   $9y^2 - 2xy = 9$

5. $\dfrac{y}{5} = \dfrac{(2x-5)}{3}$
   $2(x+y) = xy - 3$

6. $2x^2 + y^2 = 11$
   $\dfrac{(y+3)}{3} = \dfrac{(y-x)}{2}$

19.3

The general quadratic equation can be written:

$$ax^2 + bx + c = 0$$

Following the process of completing the square:

$$x^2 + \frac{b}{a} \cdot x = -\frac{c}{a}$$

$$x^2 + \frac{b}{a} \cdot x + \left(\frac{b}{2a}\right)^2 = \frac{b^2}{4a^2} - \frac{c}{a}$$

$$\left(x + \frac{b}{2a}\right)^2 = \frac{b^2 - 4ac}{4a^2}$$

$$\left(x + \frac{b}{2a}\right) = \frac{\pm\sqrt{b^2 - 4ac}}{2a}$$

$$x = \frac{-b \pm \sqrt{b^2 - 4ac}}{2a}$$

Solve the equation:

$$4x^2 + 12x + 7 = 0$$

correct to two decimal places. Applying the formula $a = 4$, $b = 12$, $c = 7$:

$$x = \frac{-12 \pm \sqrt{144 - 112}}{8}$$

$$x = \frac{-12 \pm \sqrt{32}}{8}$$

$$= \frac{\overset{-3}{-\cancel{12}} \pm \overset{}{\cancel{4}}\sqrt{2}}{\underset{2}{\cancel{8}}} = \frac{-3 \pm 1 \cdot 414}{2}$$

$$x = \frac{-4 \cdot 414}{2} \quad \text{or} \quad \frac{-1 \cdot 586}{2}$$

$$x = -2 \cdot 21 \quad \text{or} \quad -0 \cdot 79$$

This led to two roots, one taking the square root with positive sign before it, and the other with negative sign.

The portion under the root is called the discriminant, and the nature of the roots fall into three classes:

(a) (i) $b^2 - 4ac$, a complete square, gives two real unequal rational roots, and the original equation will solve into two brackets by inspection.

(ii) $b^2 - 4ac > 0$ (greater than zero, i.e. positive), but not a complete square, gives two real unequal irrational roots.

(b) $b^2 - 4ac = 0$ gives two real coincident rational roots.

(c) $b^2 - 4ac < 0$ (less than zero, i.e. negative) gives imaginary or unreal roots.

We have found that $y = ax^2 + bx + c$ is a parabola and its y values have the same sign as a, except between the roots. It crosses the x-axis ($y = 0$) at the points which are the solutions of the equations $y = ax^2 + bx + c$ and $y = 0$, i.e. $ax^2 + bx + c = 0$. The three cases are shown geometrically.

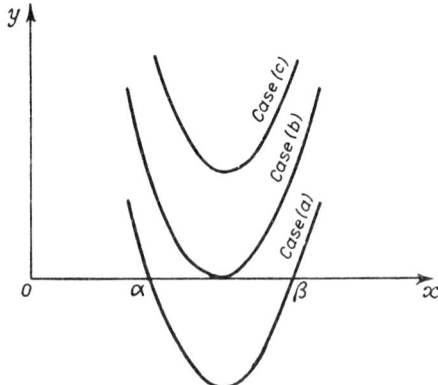

In case (a) we shall call the roots α and β. In case (b), $\alpha = \beta$. If we are given two roots α and β and wish to find the equation we work backwards from our solutions of quadratics by inspection into two brackets:

$$x = \alpha \quad \text{or} \ x = \beta$$

$$(x - \alpha) = 0 \quad \text{or} \ (x - \beta) = 0$$

$$(x - \alpha)(x - \beta) = 0$$

$$x^2 - (\alpha + \beta)x + \alpha\beta = 0$$

Obtain an equation whose roots are $+2$ and -3:

$$x = +2 \quad \text{or} \quad x = -3$$
$$(x-2)(x+3) = 0$$
$$x^2+x-6 = 0$$

Exercise 19c. Solve the following equations correct to two decimal places:

1. $x^2+x-4 = 0$
2. $x^2-3x-2 = 0$
3. $2x^2+5x+3 = 0$
4. $3x^2+2x = 2$
5. $4l^2+7l-6 = 0$
6. $4-3x^2 = 2x$

Obtain the quadratics with the following roots:

7. 1 and -4
8. -3 and -2
9. 3 and 2
10. 3 and -2
11. $\frac{1}{2}$ and $\frac{3}{4}$
12. $\frac{7}{8}$ and zero

Use the discriminant to find the nature of the roots of:

13. $x^2-4x+3 = 0$
14. $x^2-4x+2 = 0$
15. $x^2-4x+4 = 0$
16. $x^2-4x+5 = 0$
17. $x^2+x+1 = 0$
18. $x^2+x = 1$

19.4

The general equation $ax^2+bx+c = 0$ has two roots α and β. As we found in the last section they form the equation $(x-\alpha)(x-\beta) = 0$, which gives

$$x^2-(\alpha+\beta)x+\alpha\beta = 0$$

Dividing the general equation by a gives

$$x^2+\frac{b}{a}.x+\frac{c}{a} = 0$$

These are the same equation, so by comparison of terms

$$\alpha+\beta = -\frac{b}{a} \quad \text{and} \quad \alpha\beta = \frac{c}{a}$$

We can now form other equations whose roots bear a relationship with the original roots. The new equations must be reduced to terms involving $(\alpha+\beta)$ or $\alpha\beta$, as the object is to find the new equation without actually finding α and β. In fact if the equation has a negative discriminant we will be unable to find the unreal roots, although their sum and product are always real. Let us consider some terms in α and β which reduce to our sum and product:

(a)
$$\begin{aligned} \alpha^2+\beta^2 &= \alpha^2+2\alpha\beta+\beta^2-2\alpha\beta \\ &= (\alpha+\beta)^2-2\alpha\beta \end{aligned}$$

(b)
$$\begin{aligned} (\alpha-\beta)^2 &= \alpha^2-2\alpha\beta+\beta^2 \\ &= \alpha^2+2\alpha\beta+\beta^2-4\alpha\beta \\ &= (\alpha+\beta)^2-4\alpha\beta \end{aligned}$$

(c)
$$\frac{1}{\alpha^2}+\frac{1}{\beta^2} = \frac{\beta^2+\alpha^2}{\alpha^2\beta^2} = \frac{\alpha^2+\beta^2}{(\alpha\beta)^2}$$

the numerator being type (a).

(d)
$$\begin{aligned} \alpha^3+\beta^3 &= (\alpha+\beta)(\alpha^2-\alpha\beta+\beta^2) \\ &= (\alpha+\beta)(\underline{\alpha^2+\beta^2}-\alpha\beta) \end{aligned}$$

the underlined portion being type (a) again.

Type (a) is the most important, the other three give some indication of other types.

The equation $ax^2+bx+c=0$ has roots α and β. Form an equation with the roots $\alpha(1+\beta)$ and $\beta(1+\alpha)$.

The new equation has roots

$$x = \alpha(1+\beta) \quad \text{and} \quad x = \beta(1+\alpha)$$

i.e.
$$\{x-\alpha(1+\beta)\}\{x-\beta(1+\alpha)\} = 0$$

$$\therefore x^2-x(\alpha+\alpha\beta+\beta+\alpha\beta)+\alpha\beta(1+\beta)(1+\alpha) = 0$$

$$x^2-x(\alpha+\beta+2\alpha\beta)+\alpha\beta(1+\alpha+\beta+\alpha\beta) = 0$$

but
$$\alpha+\beta = -\frac{b}{a} \quad \text{and} \quad \alpha\beta = \frac{c}{a}$$

$$x^2-x\left(-\frac{b}{a}+\frac{2c}{a}\right)+\frac{c}{a}\left(1-\frac{b}{a}+\frac{c}{a}\right) = 0$$

Multiplying through by a^2

$$a^2x^2+a(b-2c)x+c(a-b+c) = 0$$

This is the required equation.

Exercise 19d. (Equations should be cleared of fractions.)

1. The equation $ax^2+bx+c=0$ has roots α and β. Form an equation with roots $\dfrac{1}{\alpha}$ and $\dfrac{1}{\beta}$.

2. The equation $ax^2 + bx + c = 0$ has roots α and β. Form an equation with roots $-\alpha$ and $-\beta$.

3. Find in terms of $(\alpha + \beta)$ and $\alpha\beta$: (a) $\alpha - \beta$, (b) $\alpha^2\beta + \alpha\beta^2$, (c) $\dfrac{1}{\alpha^3} + \dfrac{1}{\beta^3}$.

4. The equation $px^2 + qx + r = 0$ has roots α and β. Form the equations with roots $\dfrac{\alpha}{\beta}$ and $\dfrac{\beta}{\alpha}$.

5. Find the relationship between the coefficients of $px^2 + qx + r = 0$ if one root is four times the other.

6. The roots of the equation $lx^2 + mx + n = 0$ are α and β. Form an equation whose roots are $\alpha\beta^3$ and $\alpha^3\beta$.

7. The roots of the equation $x^2 - 5x + 5 = 0$ are α and β. Form an equation with roots α^2 and β^2.

8. The roots of the equation $x^2 + 2x + 3 = 0$ are α and β. Form the equation with roots $(\alpha - \beta)$ and $(\beta - \alpha)$.

20. Further Integration

20.1

To integrate the trigonometrical functions we must consider the reverse of differentiation.

$$\int \cos x \, . \, dx = \sin x + c$$

$$\int \sin x \, . \, dx = -\int -\sin x \, . \, dx$$

$$= -(\cos x) + c = -\cos x + c$$

$$\int \cos ax \, . \, dx = \frac{1}{a} \int a \, . \, \cos ax \, . \, dx$$

$$= \frac{1}{a} \, . \, \sin ax + c = \frac{\sin ax}{a} + c$$

The integration of the other four ratios is beyond the scope of this book.

H

Let us take an example of integration between limits:

$$\int_0^{\pi/6} \cos 3\theta \cdot d\theta = \left[\frac{\sin 3\theta}{3}\right]_0^{\pi/6} = \frac{1}{3}\left(\left[\sin\frac{\pi}{2}\right] - [\sin 0]\right)$$

$$= \frac{1}{3}(1 - 0) = \frac{1}{3} \text{ sq. units}$$

Exercise 20a

1. $\int \sin x° \cdot dx$

2. $\int \sin 4\theta \cdot d\theta$

3. $\int \cos x° \cdot dx$

4. $\int \cos 5x \cdot dx$

5. $\int_0^{\pi/4} \sin \theta \cdot d\theta$

6. $\int_0^{\pi/4} \cos x \cdot dx$

7. $\int_0^{\pi/3} 1 + \cos x \cdot dx$

8. $\int_0^{2\pi/3} \sin x \cdot dx$

9. Find the volume of revolution of the curve $y = \sqrt{\sin x}$, between 0 and π, through 360° about the x-axis.

10. Find the volume of revolution of the curve $y = \sec x$, between 0 and $\frac{\pi}{4}$, through 360° about the x-axis.

20.2

If we are given certain conditions we can find the value of the constant of integration.

A particle moves with constant acceleration a, starting with velocity u, when distance $s = 0$ and time $t = 0$

$$\frac{d^2 s}{dt^2} = a \qquad (1)$$

i.e.

$$\frac{d\left(\frac{ds}{dt}\right)}{dt} = a$$

$$\therefore \frac{ds}{dt} = \int a \cdot dt$$

$$\therefore \frac{ds}{dt} = at + c \qquad (2)$$

$\frac{ds}{dt}$ is the velocity, and when $t = 0$, we are told velocity $= u$, i.e.

$$u = a \times 0 + c$$

$$c = u$$

Resubstituting in (2):

$$\frac{ds}{dt} = at + u$$

$$\therefore s = \int at + u \,.\, dt$$

$$s = \tfrac{1}{2}at^2 + ut + k$$

When $t = 0$ we are told $s = 0$

$$\therefore k = 0$$

$$\therefore s = ut + \tfrac{1}{2}at^2 \qquad (3)$$

If we use the notation of section 11.1

$$\ddot{s} = a$$

$$\dot{s} = \int a \,.\, dt = at + c$$

$\dot{s}$ is the velocity at any instant (v), i.e.

$$v = u + at$$

$$s = \int u + at \,.\, dt$$

$$s = ut + \tfrac{1}{2}at^2$$

Exercise 20b

1. A particle moves with a velocity equal to $3t^2 + 2t$ in m/s, where t is the time in seconds. If it starts at the point A, find the distance from A after 3 seconds.

2. A particle has an acceleration 32 m/s^2 and starts from rest. What is its velocity after 3 s and how far has it moved?

3. A particle has an acceleration of $2t$ in m/s^2, where t is the time in seconds. It starts with a velocity of 2 m/s from A. What is its velocity and distance from A after 2 seconds?

4. If the velocity of a particle is $\dot{s} = 8 - t^3$ in m/s and it starts so that $s = 0$ when $t = 0$, find the distance from the starting point when the velocity is zero for an instant.

5. The velocity of a particle is $4t^3 + 3t^2$ in m/s and it starts $+2$ m from the point A when $t = 0$. What is its acceleration and distance from point A after 4 s?

6. A particle moves so that its velocity is given by $18t - t^2$ in m/s, where t is the time in seconds. If $s = 0$ when $t = 0$, find the distance from the starting point when the acceleration is zero.

20.3

In section 13.9 we integrated by reversing the single step function of a function differentiation. Since this was demonstrated in the beginning using a substitution we can also integrate by substitution for the terms within the brackets.

$$\frac{dy}{dx} = \frac{x}{(x^2+3)^2}$$

$$y = \int \frac{x \cdot dx}{(x^2+3)^2}$$

Let

$$u = x^2+3$$

$$\frac{du}{dx} = 2x$$

i.e.

$$x \cdot dx = \frac{du}{2}$$

by substitution:

$$y = \int \frac{1}{u^2} \cdot \frac{du}{2} = \tfrac{1}{2} \int u^{-2} \cdot du$$

$$= \frac{-1}{2u} + c$$

by substitution for u:

$$y = \frac{-1}{2(x^2+3)} + c$$

20.4

When we have a definite integral there is no need to re-substitute, but the limits have to be changed to the new variable:

$$y = \int_0^1 \frac{x \cdot dx}{(x^2+3)^2}$$

Let

$$u = x^2+3$$

$$\frac{du}{dx} = 2x$$

Upper limit, when $x = 1$: $u = 4$

Lower limit, when $x = 0$: $u = 3$

$$y = \tfrac{1}{2} \int_{3}^{4} u^{-2}.\mathrm{d}u$$

$$y = \frac{1}{2}\left[-\frac{1}{u} \right] = \tfrac{1}{2}([-\tfrac{1}{4}]-[-\tfrac{1}{3}])$$

i.e. $y = \tfrac{1}{24}$

Exercise 20c. Find by substitution:

1. $\int (x^2+3)^2.x.\mathrm{d}x$ (subs. $u = x^2+3$)

2. $\displaystyle\int_{0}^{1} \frac{x^3.\mathrm{d}x}{(x^4+1)^3}$ (subs. $t = x^4+1$)

3. $\displaystyle\int \frac{(2x^3+1).\mathrm{d}x}{(x^4+2x)^3}$ (subs. $u = x^4+2x$)

4. $\displaystyle\int_{-1}^{2} \frac{\mathrm{d}x}{(2x+3)^2}$

5. $\displaystyle\int_{0}^{2} \sqrt{x+2}.\mathrm{d}x$

6. $\displaystyle\int_{0}^{3} \frac{x.\mathrm{d}x}{\sqrt{1+x}}$ (subs. $t^2 = 1+x$)

---

21. Locus Problems; the Circle; Curve Tracing

21.1

The locus of a point is the path it traces out whilst obeying certain conditions.

In Euclidian geometry we found that if a point P moves so that it is an equal distance from fixed points A and B, its locus is the perpendicular bisector of A and B. In co-ordinate geometry let us establish the points $A(h,k)$; $B(p,q)$ and $P(x,y)$. An equation which connects x and y is the locus of P. The condition is that $PA = PB$, i.e. $PA^2 = PB^2$

$$(x-h)^2+(y-k)^2 = (x-p)^2+(y-q)^2$$

$$x^2-2hx+h^2+y^2-2ky+k^2 = x^2-2px+p^2+y^2-2qy+q^2$$

$$2y(q-k) = 2x(h-p)+p^2+q^2-h^2-k^2$$

Since A and B are fixed points h, k, p and q are constants and so we have a first-degree equation in x and y. It is therefore a straight line.

Thus $$y = \frac{(h-p)}{(q-k)}x+\frac{p^2+q^2-h^2-k^2}{2.(q-k)}$$

is the locus of point P.

The slope is $$-\frac{(h-p)}{(k-q)}$$

The slope of AB is $$\frac{(k-q)}{(h-p)}$$

and therefore the locus is perpendicular to AB. It must bisect AB because the mid-point of the line AB is

$$\left(\frac{h+p}{2}, \frac{k+q}{2}\right)$$

which satisfies the locus of P.

Exercise 21a. Find the locus of the point $P(x,y)$ that obeys the following conditions:

1. An equal distance from $(2,1)$ and $(2,4)$.
2. An equal distance from $(3,-1)$ and $(-2,2)$.
3. A fixed distance r from the origin.
4. A fixed distance 3 from the point $(-3,-2)$.
5. Twice as far from the point $(2,1)$ as it is from $(-3,2)$.
6. The line from P to $(5,4)$ is perpendicular to the line from P to the origin.
7. Lies on the line through $(2,-3)$ and $(-5,-7)$.
8. Lies at an equal distance from the line $x = -1$ and the point $(1,0)$.

21.2

If a point P moves in a plane, so that it is a constant distance from a fixed point A the locus of P is a circle. The fixed point A is the centre of the circle which in co-ordinate geometry we shall call $C(\alpha,\beta)$ and the fixed distance is the radius a. The point P will be (x,y) and by Pythagoras' theorem:

$$(x-\alpha)^2+(y-\beta)^2 = a^2$$
$$x^2+y^2-2\alpha x-2\beta y-a^2+\alpha^2+\beta^2 = 0$$

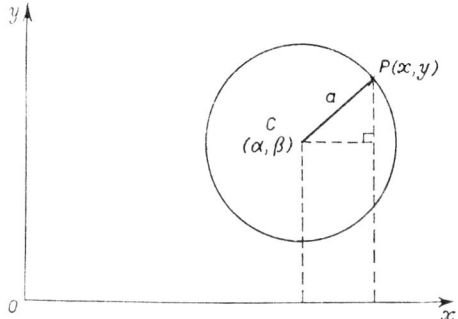

This is a second-degree equation with the coefficients of x^2 and y^2 unity and there is no xy term. Thus if a second-degree equation has no xy term, and the same coefficients of x^2 and y^2, we can divide through by that coefficient and write the equation in the form $x^2+y^2+2gx+2fy+c=0$.

Comparing the two equations

$$g = -\alpha; \quad f = -\beta \quad \text{and} \quad -a^2+\alpha^2+\beta^2 = c$$

Thus $\alpha = -g$ and $\beta = -f$. Substituting these values in the final equation $a = \sqrt{g^2+f^2-c}$.

The centre is $(-g, -f)$ and radius $\sqrt{g^2+f^2-c}$. Find the centre and radius of the circle

$$2x^2+2y^2-2x+2y-1 = 0$$
$$\therefore x^2+y^2-x+y-\tfrac{1}{2} = 0$$
$$x^2+y^2+2(-\tfrac{1}{2})x+2(\tfrac{1}{2})y-\tfrac{1}{2} = 0$$

Thus, $g = -\tfrac{1}{2}, f = +\tfrac{1}{2}$, and $c = -\tfrac{1}{2}$. The centre is $(\tfrac{1}{2}, -\tfrac{1}{2})$. The radius is $\sqrt{\tfrac{1}{4}+\tfrac{1}{4}-(-\tfrac{1}{2})} = 1$.

If the centre is at the origin, $\alpha = \beta = 0$, and the equation becomes $x^2+y^2 = a^2$.

Exercise 21b

1. Find the equation of the circle with centre $(3,3)$ and radius 3.
2. Find the equation of the circle with centre $(4, -1)$ and radius 2.
3. Find the centre and radius of the circle $x^2+y^2-25 = 0$.
4. Find the centre and radius of the circle $x^2+y^2-4x-2y-11 =0$.

5. A circle has centre $(\frac{1}{2}, \frac{3}{4})$ and radius 2. What is its equation?

6. $x^2 + y^2 + 4x - 6y - 3 = 0$ is a second-degree equation. Why is it a circle and what is its centre and radius?

7. What is the equation of a circle centre $(-2, -4)$ with radius 5?

8. Find the centre and radius of the circle
$$9x^2 + 9y^2 - 12x - 6y - 13 = 0.$$

21.3

To find the equation of the tangent to the circle $x^2 + y^2 = a^2$ (the origin being the centre) at the point $P_1 (x_1, y_1)$, we have only to find its slope at this point. This is because the tangent is a straight line and passes through the point $P_1(x_1, y_1)$, so by section 16.2 its equation is

$$y - y_1 = m(x - x_1)$$

By section 4.2 the slope of a tangent of a curve is given by its first derivative.

The circle is

$$y^2 = a^2 - x^2$$

$$2y.\frac{dy}{dx} = -2x \quad \text{i.e.} \quad \frac{dy}{dx} = -\frac{x}{y}$$

The slope at $P_1(x_1, y_1)$ is $-\dfrac{x_1}{y_1}$.

The equation of the tangent is

$$y - y_1 = -\frac{x_1}{y_1}(x - x_1)$$

i.e.
$$xx_1 + yy = y_1^2 + x_1^2$$

but the point $P_1(x_1, y_1)$ also lies on the circle, so

$$y_1^2 + x_1^2 = a^2$$

Therefore the equation of the tangent at $P_1(x_1, y_1)$ is $xx_1 + yy_1 = a^2$.

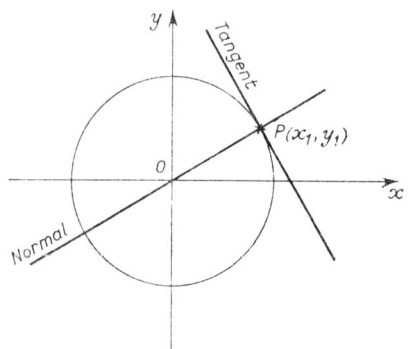

21.4

The normal is perpendicular to the tangent and so its slope is $+\dfrac{y_1}{x_1}$
by the result of section 8.4.

Since it passes through $P_1(x_1, y_1)$ its equation is

$$y - y_1 = \frac{y_1}{x_1}(x - x_1)$$

i.e.

$$x_1 y - x_1 y_1 = y_1 x - x_1 y_1$$

The equation to the normal at $P_1(x_1, y_1)$ is $y_1 x = x_1 y$.

Since it is satisfied by the point $(0,0)$ the normal passes through the origin, which is the centre of the circle.

Exercise 21c

1. Find the tangent to the circle $x^2 + y^2 = 2$ at the point $(-1, -1)$.
2. Find the normal to the circle $x^2 + y^2 = 34$ at the point $(-3, 5)$.
3. Find the equation of the tangents of the circle $x^2 + y^2 - 10 = 0$ at the points where $x = 1$.
4. Find the equations of the normals of the circle $x^2 + y^2 - 169 = 0$ at the points where $y = 12$.
5. Find the equations of the tangents of the circle $x^2 + y^2 = 9$ at the points $(-3, 0)$ and $(0, 3)$. Show that they are perpendicular to each other.
6. The normal at the point where $x = 3$ is $x + 6y = 0$. What is the equation of the circle? (the centre is at the origin).

21.5

To trace or sketch a curve we are trying to show the general shape of the curve with its outstanding features. The following points give an outline of what is needed, and they should be shown if they can be found easily.

(i) Symmetry $y^2 = 8x$ is symmetrical about the x-axis $(y = 0)$ because each value of x gives two values of y of the same size but different signs, $y = \pm\sqrt{8x}$.

$y = x^2 + 4x + 3$ on completing the square gives

$$y = (x+2)^2 - 1$$

i.e. $x = -2 \pm \sqrt{y+1}$, which is symmetrical about the line $x = -2$.

(ii) The points where the graph cuts the two axes. $y^2 = 8x$ cuts the x-axis $(y = 0)$ at $x = 0$, and, of course, the y-axis at $y = 0$.

The other curve cuts the y-axis at $y = 3$. For the other axis $0 = (x+3)(x+1)$, i.e. $x = -1$ or -3. Note that this would not have been found if the expression could not have been solved by putting into brackets.

(iii) Parts of the system where the graph is unreal. If $y^2 = 8x$ we can only find y if the right-hand side is positive or zero, i.e. x is positive or zero. The curve does not exist for negative values of x.

For the second curve $y + 1 = (x+2)^2$. Here the left-hand side must be positive or zero to find the square root, so y must be greater than or equal to -1, which is written $y \geqslant -1$.

(iv) Maxima or minima. When $y^2 = 8x$, we have $2y \cdot \dfrac{dy}{dx} = 8$, i.e. no max. or min.

When $\qquad y = (x+2)^2 - 1$

$$\frac{dy}{dx} = 2(x+2) \quad \text{and} \quad \frac{d^2y}{dx^2} = 2$$

For a turning point $\dfrac{dy}{dx} = 0$, i.e. $x = -2$, then $y = -1$, and it is a minimum as the second derivative is positive.

(v) The large negative and positive values of x or y. The first curve flattens out for large values. The second equation approximates to $y = x^2$ for large values.

(vi) The points where $x = 1$ and $y = 1$. For $y^2 = 8x$ when $x = 1$, $y = \pm \sqrt{8}$; when $y = 1$, $x = \frac{1}{8}$.

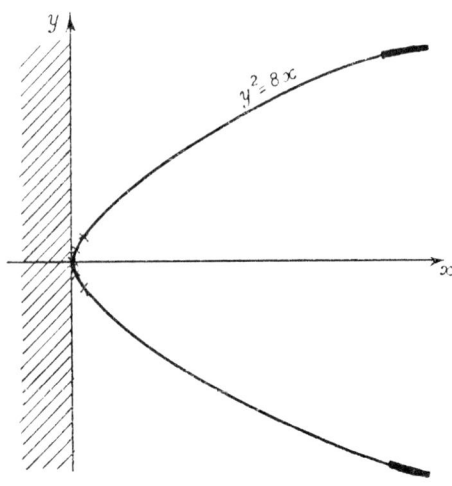

For $y = x^2 + 4x + 3$, when $x = 1$, $y = 8$. When $y = 1$ the value would take too much working to find.

(vii) Small values of x and y. For the curve $y^2 = 8x$, the x values become very small when y is less than 1.

The curve $y = x^2 + 4x + 3$ approximates to $y = 4x + 3$ when x is small. Notice that the curve is not drawn strictly to scale.

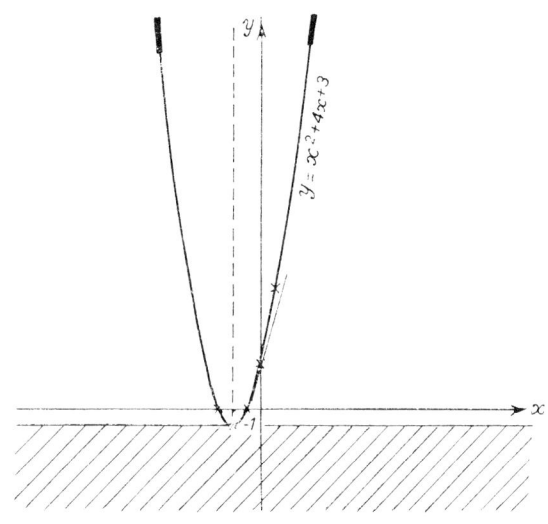

Exercise 21d. Trace the following curves:

1. $y = x^2 - 4x + 5$.
2. $y = x(x+1)(x-1)$.
3. $y^2 + 7y + 12 = x$.
4. $xy = 1$.
5. $y = x^3 + 3$.
6. $y = (x-1)(x-2)(x+2)$.
7. $y^2 = (x-1)(x-2)(x+2)$, i.e. $y = \pm \sqrt{(x-1)(x-2)(x+2)}$, so consider what happens to curve of number 6 when root is taken.
8. $y = [(x-1)(x-2)(x+2)]^2$.

PART TWO

Applied Mathematics

22. Linear Kinematics

22.1

The results of this chapter apply to motion in a straight line. In Chapter 11 we found that the instantaneous velocity of a body which has moved distance s in time t is given by $\dfrac{ds}{dt}$. It can be measured in m/s or km/h and a useful relationship is

$$18 \text{ km/h} = 5 \text{ m/s}$$

A knot is a nautical mile per hour where a nautical mile can be taken as 1850 m. The nautical mile is the international unit of length for sea or air navigation.

If we plot a graph of distance against time the slope at any point will be $\dfrac{ds}{dt}$, the velocity. Thus if the graph is a straight line it will have a constant slope and will represent a constant or uniform velocity. If the graph is parallel to the t-axis in one portion it means that the object must be stationary, because its distance from the starting point remains constant.

22.2

The acceleration or rate of change of velocity was found to be $\dfrac{dv}{dt}$ at any instant. Other forms are $\dfrac{d^2s}{dt^2}$ or $v.\dfrac{dv}{ds}$. Since it is only likely to occur for up to a minute in everyday cases it is measured in m/s^2 or mm/s^2. A negative acceleration is called a retardation.

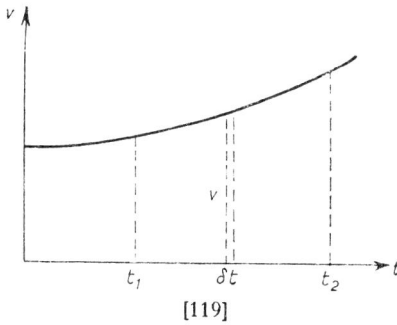

If we plot a graph of velocity against time the slope will represent the acceleration for it is $\dfrac{dv}{dt}$.

The area under the graph of the small element is $v.\delta t$. In the limit the area is $\int_{t_1}^{t_2} v.dt$ and therefore must represent the distance travelled between times t_1 and t_2. If the graph is a straight line it represents a constant or uniform acceleration. If it is parallel to the t-axis the velocity is constant. If it runs along the t-axis the velocity is zero.

Exercise 22a

1. A ship is travelling at 30 knots. How many metres is it covering every second?

2. A particle starts at the point A and moves such that s in metres $= 8t - t^2$, where t is the time in seconds. Construct a table showing the distance from A (s) at the end of each second up to 10 seconds. Plot a graph of distance against time. From it find when the particle is at rest for an instant. A second particle starts from the same point A at the same time with a velocity of 3 m/s in the same direction. At what time do the particles meet and how far are they from the start?

3. A particle moves at 3 m/s in a straight line from A towards B for 8 seconds and is then stationary for 5 s. It returns to the starting point A at 4 m/s. A second particle starts at B, which is 30 m from A, 6 s after the other particle has first started and moves towards A at $1\frac{1}{2}$ m/s. Plot a distance–time graph and find at what times the particles are in the same positions.

4. A motor car accelerates from rest to 60 km/h in 5 seconds. What is its average acceleration in m/s²?

5. A particle accelerates at 3 m/s² for 6 seconds and is then retarded at 2 m/s² until it is again at rest. Plot a velocity–time graph, find the maximum velocity and the distance travelled. (Note the shape of the area under the graph.)

6. A particle accelerates at 4 m/s² from rest until it reaches 44 m/s. It then travels at this speed for 22 seconds and is brought to rest in 22 s. Plot a velocity–time graph and find the distance travelled.

22.3

A body starts with an initial velocity u at the point $s = 0$, when $t = 0$, and moves with constant acceleration a. If it reaches a final velocity v after time t and has moved a net distance s, we can represent it as a velocity–time graph. It is a straight line because the acceleration is constant.

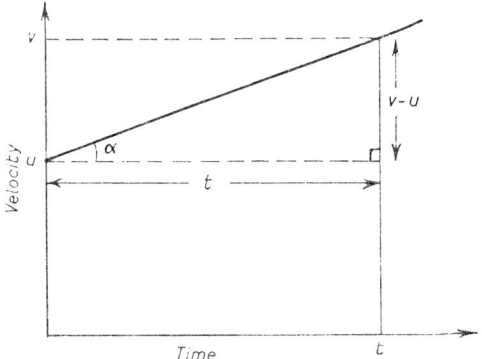

The slope $\tan\alpha = \dfrac{v-u}{t}$ but the slope of a velocity–time graph is the acceleration a, i.e.

$$a = \frac{v-u}{t}$$

or
$$v = u+at \tag{1}$$

The area under the graph is equal to the distance s. We start by treating it as a rectangle of area ut and a triangle of area $\frac{1}{2}(v-u).t$. But from (1)

$$(v-u) = at$$
$$\therefore s = ut + \tfrac{1}{2}at^2 \tag{2}$$

Equations (1) and (2) have been established in section 20.2 by the use of calculus.

If we treat the area under the graph as a trapezium we obtain

$$s = \frac{(u+v)}{2}t \tag{3}$$

From equation (1) $t = \dfrac{(v-u)}{a}$ and if it is substituted in (3) we obtain

$$s = \frac{(u+v)(v-u)}{2a}$$

i.e.
$$2as = v^2 - u^2$$

$$v^2 = u^2 + 2as \tag{4}$$

ı

From equation (1) $u = v - at$ and if this is substituted in (3)

$$s = \frac{(v - at + v)t}{2}$$

$$s = vt - \tfrac{1}{2}at^2 \tag{5}$$

Notice that we have five quantities a, v, u, t, and s and each of our equations leaves out one. Therefore if we are given three of the quantities we can find the other two.

A particle increased its velocity from 54 km/h to 90 km/h in 0·1 km. What was the constant acceleration and how long did it take? Let us list the known quantities converting them all into suitable units.

Initial velocity: $u = 54$ km/h $= 15$ m/s
Final velocity: $v = 90$ km/h $= 25$ m/s
$\qquad\qquad s = 0·1$ km $= 100$ m

It is easier to find t first, using (3):

$$100 = \frac{(40)}{2}.t \quad \text{i.e. } t = 5 \text{ s}$$

then from (1): $\qquad 25 = 15 + a \times 5$

$$a = \frac{10}{5} = 2 \text{ m/s}^2$$

Exercise 22b (All at constant acceleration.)

1. A particle increases its velocity from 72 km/h to 108 km/h in 9 s. What is its acceleration and the distance it has travelled?
2. A ship increases its velocity from 5 knots to 25 knots in 10 min. What is its acceleration and the distance it has travelled? (Give answers in terms of hours and nautical miles.)
3. A particle is accelerated at 4 m/s² over 100 m in 5 seconds. What are its initial and final velocities?
4. A particle accelerates at 3 m/s² from 33 to 99 m/s. What distance has it travelled and how long does it take?
5. A particle accelerates at 2 m/s² over 150 m to finish with a speed of 40 m/s. What was the initial velocity and how long does it take?
6. A particle starts with a velocity of 20 m/s and travels 90 m in 3 seconds. What is the acceleration and the final velocity?

22.4

A very important application of a body moving in a straight line with constant acceleration is that of a free falling body. Strictly speaking this should be in a vacuum, but for many problems of falling in air the resistance of the air can be ignored. Under these conditions all bodies fall with an acceleration (g) that can be taken as 9·8 m/s², or 9 800 mm/s².

A body falls from rest for 10 seconds. Find its final velocity and the distance fallen (in metres).

$$u = 0, \quad g = 9\cdot8 \text{ m/s}^2, \quad t = 10 \text{ s}$$
$$v = u + gt = 0 + 9\cdot8 \times 10 = 98 \text{ m/s}$$
$$s = ut + \tfrac{1}{2}gt^2 = 0 + \tfrac{1}{2}9\cdot8 \times 10^2 = 490 \text{ m}$$

Here we have taken the distance downwards as positive, making the acceleration positive, and a downwards velocity is positive.

A body is projected upwards with a speed of 20 m/s. Here we shall take the upwards measurement as positive, thus

$$a = -9\cdot8 \text{ m/s}^2$$

(i) To what height does it rise?

$$u = +20 \text{ m/s}, \quad v = 0, \quad a = -9\cdot8 \text{ m/s}^2$$
$$v^2 = u^2 + 2as$$
$$0 = 400 - 19\cdot6s$$
$$s = \frac{400}{19\cdot6} = 20\cdot4 \text{ m (correct to 3 s.f.)}$$

(ii) What is its velocity when it returns to the starting point?

$$u = 20, \quad a = -9\cdot8, \quad s = 0$$
$$v^2 = 400 - 19\cdot6 \times 0$$
$$v = \pm20 \text{ m/s}$$

By our notation the $+20$ m/s is the starting upward velocity. The returning velocity is -20 m/s, i.e. 20 m/s downwards. It will be seen that a body always returns with its starting speed, but in the opposite direction. Notice that the distance measured is the net distance from the origin. A negative distance means that it is below the starting point if that is possible, i.e. near the edge of a cliff.

Exercise 22c

1. A body is dropped from the top of a building. How far has it fallen in metres at the end of 1 second, 2 s, and 5 s?

2. What is the velocity of a dropped free-falling body after (a) 0·2 s in mm/s, (b) 3 s in m/s, (c) 4 s in km/h?

3. A particle is projected downwards with a velocity of 40 m/s. How far has it travelled after 5 seconds? How far does it travel in the next 5 seconds?

4. An object is projected upwards with a velocity of 245 m/s. Calculate its position after 10 seconds. Find the total time that has elapsed when it has returned to that position again.

5. A cliff is 100 m high. A particle is projected upwards from the top with a velocity of 40 m/s. How long does it take to reach the sea?

6. What is the velocity of a particle after it has fallen 14 m from rest (a) on Earth, (b) on the moon ($g = 1·62$ m/s²)?

22.5

The more difficult problems will consist of a number of parts, each of which will involve constant acceleration or velocity. They can be solved using any of the methods already described in the previous sections.

Two bodies are 165 m apart travelling towards each other. P is travelling at 6 m/s with an acceleration of 4 m/s². Q has twice the speed of P and a 2 m/s² acceleration. Find the distance P has travelled before the meeting and the time taken.

$$PQ = 165 \text{ m}$$

For P (measured towards Q):   For Q (measured towards P):

$u_p = 6$ m/s	$u = 12$ m/s
$a_p = 4$ m/s²	$a = 2$ m/s²
distance $= s_p$	distance $= 165 - s_p$
time $= t$	time $= t$

$$s = ut + \tfrac{1}{2}at^2$$

For P: $s_p = 6t + 2t^2$
For Q: $165 - s_p = 12t + t^2$

Adding: $165 = 18t + 3t^2$
$$0 = t^2 + 6t - 55$$
$$0 = (t + 11)(t - 5);$$
$$t = 5, \text{ or } -11$$

But t cannot be negative, $\therefore t = 5$ s

$$s_p = 30 + 50 = 80 \text{ m}$$

Therefore P has travelled 80 metres in 5 seconds when it meets Q.

Exercise 22d

1. A body starting from rest accelerates for 10 seconds at 2 m/s^2 and then retards at 1 m/s^2 for 10 s. If this is repeated every 20 seconds, find how far it has gone in a minute.

2. Two bodies are 120 m apart, the first being at rest and the second moving towards the first at 4 m/s. If they are both accelerating at 4 m/s^2 towards each other, how long is it before they meet?

3. Particles P and Q are 30 m apart. P is travelling towards Q at 2 m/s and is being retarded at 2 m/s^2. Q is travelling away from P at 3 m/s and is retarded at 4 m/s^2 (i.e. accelerated towards P). How far is P from its starting point when overtaken by Q, and how long has it taken?

4. A particle at rest is accelerated at a in m/s^2 for five seconds, moves with constant velocity for five seconds and is then retarded at $2a$ in m/s^2 to come to rest 300 m from the starting point. What is its acceleration and how long is it retarded?

5. A particle is moving at 20 m/s at point P and is accelerated at 3 m/s^2 until its velocity is 60 m/s. It is then retarded at 4 m/s^2 until it comes to rest at Q. Find the distance PQ and the time taken to get from P to Q.

6. A tube train has to travel one kilometre between stations. It accelerates from rest to 36 km/h in 20 seconds, and travels at this speed for $1\frac{1}{3}$ min. Find the retardation in m/s^2 to bring it to rest at the next station.

23. Composition and Resolution of Forces; Moments

23.1

A scalar quantity has magnitude only, such as mass, time or speed. A vector quantity has both magnitude and direction such as displacement, force, acceleration or velocity. We shall refer to forces but the results also apply to the other vector quantities. We shall measure forces in newtons, but if other units are met with they will not change the methods.

If we have two or more forces we must be able to compose them and find the resultant single force which will represent them. Let us start with two forces P and Q which are acting at a point A with an angle α between them. This is represented by the space diagram.

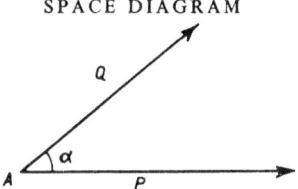

SPACE DIAGRAM

We then draw a force diagram to scale with $\overrightarrow{AB}$ to represent P and $\overrightarrow{AD}$ to represent Q.

FORCE DIAGRAM

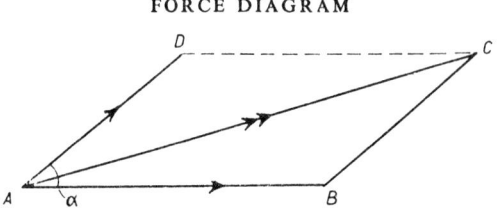

Notice that $\overrightarrow{AB}$ represents P in both magnitude and direction. We now complete the parallelogram to point C by drawing parallels to AD and AB. Then $\overrightarrow{AC}$ represents the resultant R which is shown by two arrows. Since the force diagram has to give us only the magnitude and direction of the resultant we need only construct the triangle of forces ABC. The point of action is seen to be A from the space diagram.

Forces of 2 N and 3 N act at $60°$ to each other. What is the resultant force and what angle does it make with the first force?

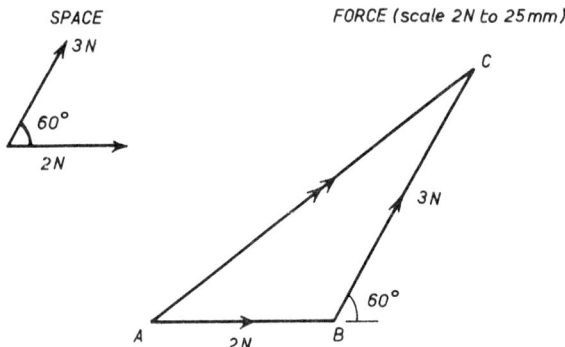

The original scale (1 N to 20 mm) gives $AC = 87$ mm, $C\hat{A}B = 37°$, and the diagram is of reasonable size giving an accuracy of 1 mm and

0·5 degrees. The resultant is 4·35 N making an angle of 37° with the first force.

Notice the result of adding two vectors can come to any value between the difference of the vectors and the sum of the vectors. In our example when the 2 N and 3 N are opposite, the resultant is 1 N, and when in the same direction, the resultant is 5 N.

Exercise 23a. Find the resultant and the angle it makes with the first vector when the following pairs have the given angle between them.

1. 2 N and 4 N, 30°.

2. 3 N and 4 N, 60°.

3. 3 N and 4 N, 90°.

4. 3 N and 4 N, 120°.

5. 6 km/h and 5 km/h, 150°.

6. 8·8 m/s and 10 m/s, 20°.

23.2

With several forces we can take a pair and obtain a resultant. This can be taken with a third force to give a resultant and the process can be continued until there is one final resultant. If $\overrightarrow{AB}$ represents P, and

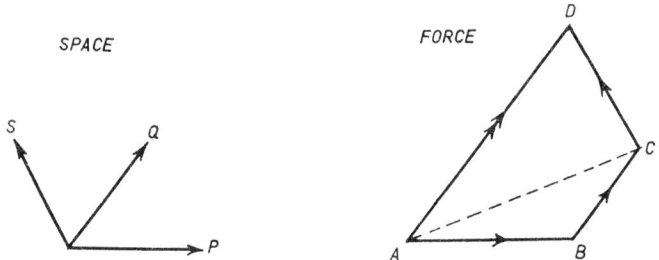

$\overrightarrow{BC}$ represents Q, the resultant is $\overrightarrow{AC}$. Then $\overrightarrow{AC}$ is taken with $\overrightarrow{CD}$, which represents S, to give a resultant $\overrightarrow{AD}$. There is no need to put in the intermediate resultant so we draw the polygon of forces $ABCD$. Note that if a different order is taken the polygon is a different shape but the point D is unchanged.

Forces 2 N and 5 N are at an angle of 67°. A force of 7 N makes an angle of 120° with the second force on the opposite side to the first. What is the resultant force?

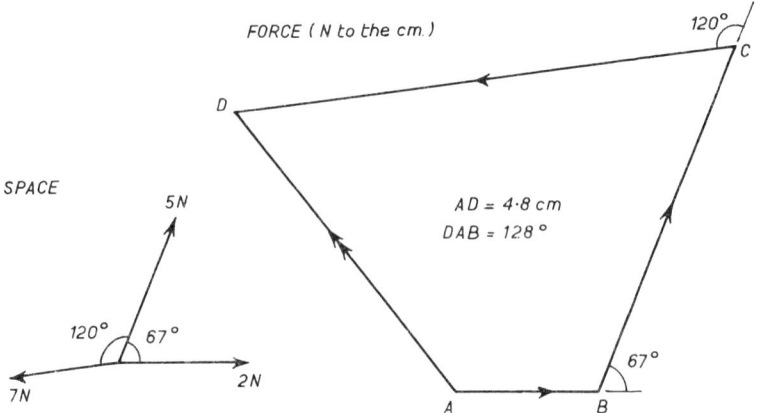

The resultant is 4·8 N making an angle 128° with the first force in the direction of the second.

To find the resultant of forces by calculation we must find out how to resolve a single force into its components.

Exercise 23b. Find the resultant and the angle it makes with the first vector, in the direction of the second, when the following act at a point:

1. A force 2 N makes an angle of 50° with a force of 5 N. A third force of 3 N makes an angle of 90° with the first force on the same side as the second.

2. As for Question 1, with third force on the opposite side to the second.

3. Forces of 3 N, 5 N, and 7 N at 120° to each other.

4. A velocity of 70 km/h is at an angle of 80° with a velocity of 50 km/h. A third velocity of 25 km/h makes an angle of 180° with the second.

5. The following forces are taken in order. 7 N makes an angle of 35° with one of 4 N, which makes an angle of 74° with 3·5 N, which makes an angle of 86° with 6·2 N.

6. The following are the angles in an anti-clockwise direction made with a force of 2 N. A 3 N force at 50°, 1 N at 110°, 5 N at 260°, and 4 N at 350°.

23.3

Two forces have been composed into a single resultant. We shall now reverse the process and split a force into two components which are at right angles to each other.

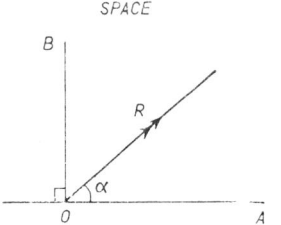

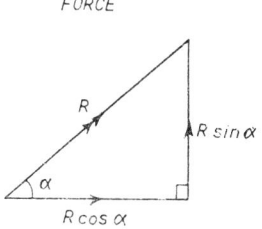

If a force R is to be resolved into two forces, one along OA and the other along OB, we consider the force diagram. The triangle of forces must have one component parallel to OA and the other parallel to OB. By trigonometry the component or resolved part along OA is $R \cos \alpha$, and that along OB is $R \sin \alpha$, where α is the angle between force R and OA.

Thus if a force of 8 N acts at O, at an angle $45°$ to a line OA, its resolute along OA is $8 . \cos 45°$ N, and perpendicular to OA it is $8 . \sin 45° $ N.

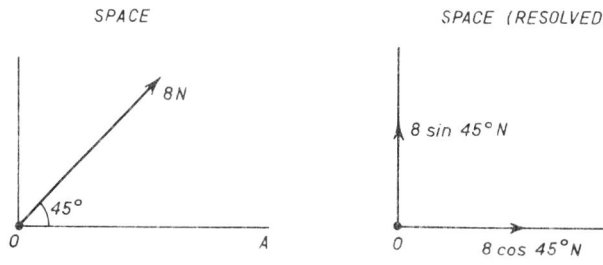

$$8 . \cos 45° = 8 \times \frac{1}{\sqrt{2}} = 8 . \frac{\sqrt{2}}{2} = 4\sqrt{2} \text{ N along } OA$$

$$8 . \sin 45° = 4\sqrt{2} \text{ N perpendicular to } OA$$

A force of 10 N makes an angle of $120°$ with a force of 6 N. What are the net forces in the direction of the first force and perpendicular to it?

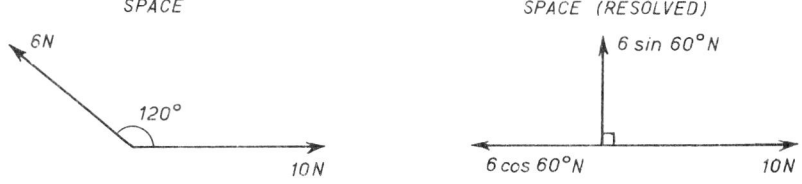

The resolved part $6 . \cos 60° = 6 \times \frac{1}{2} = 3$ N, and is to the left.

This gives a net force to the right of $10-3 = 7$ N. The 10 N has no resolute perpendicular to itself because 10.cos 90° is zero. The upward force is $6 \cdot \sin 60° = 6 \times \frac{\sqrt{3}}{1} = 3\sqrt{3}$ N.

23.4

We can now solve the resultant problems of section 23.1 by theoretical methods. Taking the example of that section we start with the space diagram.

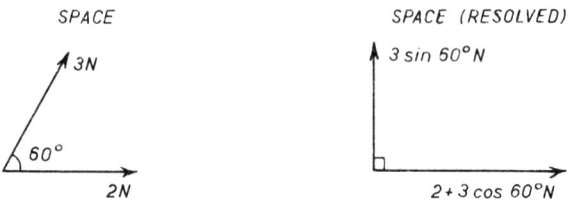

The net force to the right is $2+3 \cdot \cos 60° = 3 \cdot 5$ N. The upward force is $3 \cdot \sin 60° = 3 \times 0 \cdot 866 = 2 \cdot 598$ N. Drawing a rough force diagram:

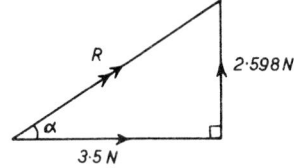

Then, by Pythagoras' theorem:

$$R^2 = 3 \cdot 5^2 + 2 \cdot 598^2$$
$$= 12 \cdot 25 + 6 \cdot 75$$
$$= 19 \cdot 00$$
$$R = 4 \cdot 36 \text{ N}$$

Also $\quad \tan \alpha = \dfrac{2 \cdot 598}{3 \cdot 5} = 0 \cdot 7423 \, ; \quad \alpha = 36° \, 35'$

The resultant of 4·36 N makes an angle of 36° 35' with the 2N.

Exercise 23c. Repeat Exercise 23a using the theoretical methods of section 23.4.

23.5

The moment of force about a point is defined as the product of the force and the perpendicular distance from the point to the line of action of the force. The most usual units are N m or N mm.

To understand the terms we can consider an object in the shape of a right-angled isosceles triangle of shorter side 1 m. A force of 1 N acts at A and 2 N acts at B as shown.

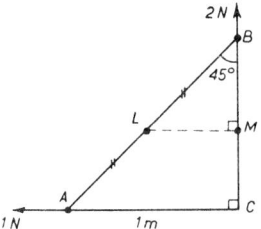

Let us consider moments about A. The first important point to note is that the force of 1 N has zero moment because it goes through A. The 2 N acts at B, and its line of action is a continuous straight line through B and C. The perpendicular distance from A to this line is 1 m. The moment about A is therefore 2 N m and is said to be anti-clockwise since it tends to turn the body round A in that way (i.e. in a direction opposite to the way the hands turn about the centre of a clock). Similarly, the moment of the 1 N about B is 1 N m in a clockwise direction.

To find the net moment about L, the mid-point of AB, we have to use geometrical or trigonometrical results to find the perpendicular distances to the lines of action. By similar triangles $LM = 0.5$ m, or by trigonometry

$$AB = \sqrt{2} \text{ m}$$

$$BL = \frac{\sqrt{2}}{2} \text{ m}$$

and
$$LM = \frac{\sqrt{2}}{2} . \sin 45° = \frac{\sqrt{2}}{2} . \frac{1}{\sqrt{2}} = 0.5 \text{ m}$$

The moment about L in an anti-clockwise direction is
$$2 \times 0.5 \text{ N m} = 1 \text{ N m}$$

The clockwise moment about L is
$$1 \times 0.5 \text{ N m} = 0.5 \text{ N m}$$

There is a net anti-clockwise moment of 0.5 N m about L. It is obvious from these results that the triangle is not in equilibrium (i.e. still).

23.6

The principle of moments states that if a body is in equilibrium the sum of the moments in a clockwise direction about any point is equal to the sum of the moments in an anti-clockwise direction about the same points.

We shall start by applying this to levers, which are supported by fulcrums. If a mass in kg of M is hung from the lever it will exert a force vertically downwards in newtons of Mg (see 26.2). This is the weight of the body and should be shown as an arrow, whereas the mass is drawn as a small square or circle. If the weight of the lever is negligible compared with the other forces acting, it is said to be a light rod. A uniform rod will have its centre of gravity in the middle of the rod; the centre of gravity being the point at which all the weight of the rod can be taken to act. When drawing a diagram we draw in all the forces acting on the lever. The support given by a fulcrum is called the reaction or thrust.

For example a uniform 2 m beam AB of mass 10 kg is supported by fulcrums under the points 0·6 m and 1·1 m from A. A mass of 5 kg is hung 0·4 m from A. (i) What are the reactions? (ii) What is the maximum mass that can be hung at the end B so that the bar is still in equilibrium?

The rule is uniform so its weight acts 1 m from A.

(i)

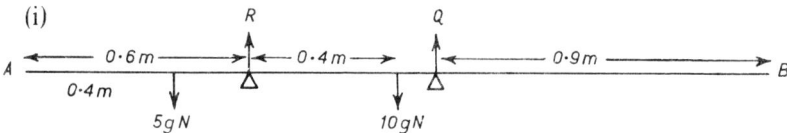

Firstly we equate the upward and downward forces:

$$R + Q = 5g + 10g \text{ N}$$

We can choose any point about which to take moments using the principle. The work is simplified by choosing a point which gives one of the unknowns zero moment. We shall take the left-hand fulcrum.

$$5g \times 0·2 + Q \times 0·5 = 10g \times 0·4$$
$$0·5Q = 4g - g = 3g$$
$$Q = 6g = 6 \times 9·8 = 58·8 \text{ N}$$

From the first equation $R = 15g - Q = 15g - 6g = 9g = 88·2$ N.

These values can be checked by taking moments about some other point, i.e. the left-hand end of the rule:

Clockwise moments $= 5g \times 0·4 + 10g \times 1·0 = 12g$ N

Anti-clockwise moments $= 9g \times 0·6 + 6g \times 1·1 = 12g$ N

The reactions are 88·2 N and 58·8 N.

(ii)

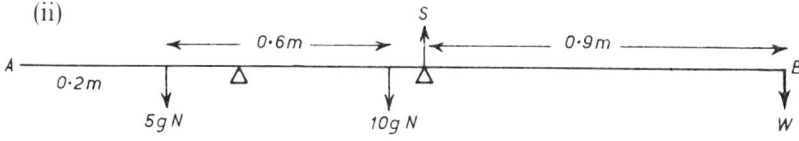

When the maximum mass is placed on the rod the left-hand fulcrum is no longer supporting any weight. As we do not have to find the reaction S we take moments about the right-hand fulcrum.

$$10g \times 0 \cdot 1 + 5g \times 0 \cdot 7 = W \times 0 \cdot 9$$
$$45g = 9W$$
$$W = 5g = 49 \text{ N}$$

The maximum mass is 5 kg, since this mass has weight $5g$ N.

Exercise 23d. (All rods in equilibrium.)

1. A light rod 2·4 m long is pivoted 0·9 m from end A. A 5 kg mass is hung 2·1 m from A and a 2·5 kg mass is hung 1·5 m from A. What mass must be placed at A and what is the reaction?

2. A uniform metre rule has a fulcrum under the 0·3 m mark. It is balanced by 0·5 kg hung under the 0·15 m mark and 0·05 kg under the 0·6 m mark. What is the mass of the rule?

3. Repeat Question 1 with a uniform rod 2·4 m long of mass 7·5 kg.

4. A uniform bar AB of length 3·6 m and mass 15 kg is balanced by fulcrums 2 m and 3 m from A. A mass of 8 kg is hung 2·4 m from A and 1 kg is hung 3·2 m from A. What are the reactions?

5. A uniform metal beam of mass 150 kg and length 4 m is supported by a fulcrum 1 m from end A and a second fulcrum which has a reaction of 588 N. How far is it from A? What is the other reaction?

6. A non-uniform bar of length 6 m and mass 40 kg has its centre of gravity 4 m from end A. It has a mass of 10 kg at 1 m from A, and 50 kg at 3·5 m from A. How far must the fulcrum be placed from end A to balance? How far is it from end A if the 10 kg and 50 kg masses are removed?

24. Equilibrium; Friction

24.1

When a body is in equilibrium we can find three unknowns by obtaining three equations in one of the following ways:

(1) Resolving in two directions and taking moments about a point.
(2) Resolving in one direction and taking moments about two points.
(3) Taking moments about three non-linear points.

Method (1) is the most useful and we shall apply it to a harder lever problem.

A non-uniform bar AB is 2 m long and has mass 10 kg. It is supported in the horizontal position by ropes at its ends. The rope at B makes an angle of $150°$ with the bar, and the rope at A makes an angle of $120°$ with the bar. Calculate the tension in each rope and the distance of the centre of gravity from A.

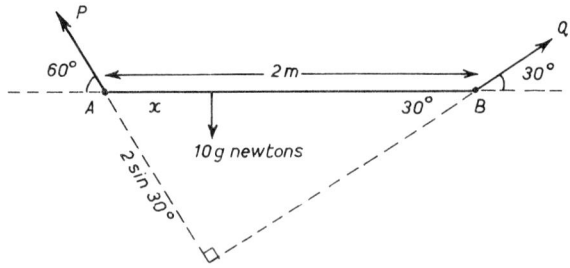

Resolving vertically:

$$10g = P.\sin 60° + Q.\sin 30°$$

$$10g = \frac{\sqrt{3}}{2}.P + \frac{Q}{2}$$

$$20g = \sqrt{3}.P + Q \qquad (1)$$

Resolving horizontally:

$$P.\cos 60° = Q.\cos 30°$$

$$\frac{P}{2} = \frac{\sqrt{3}}{2}.Q$$

$$P = \sqrt{3}.Q \qquad (2)$$

Eliminating P from (1) and (2):

$$20g = \sqrt{3}.\sqrt{3}.Q + Q$$
$$Q = 5g = 49 \text{ N}$$

From (2):

$$P = 5\sqrt{3}g = 49\sqrt{3} \text{ N} \quad \text{(check in (1))}$$

These are the most usual directions for resolution but others can be taken. In this case resolving in the direction of Q:

$$Q = 10g \cos 60° = 10g \tfrac{1}{2} = 5g \text{ N}.$$

(There is no P component, since it happens to be perpendicular to Q.)

As we have to find x it is easiest to take moments about A:

$$10g \times x = Q \times 2.\sin 30° \quad \text{but } Q = 5g$$
$$\therefore \ 10gx = 5g \times 2 \times \tfrac{1}{2}$$
$$x = 0.5 \text{ m}$$

The tensions in the ropes at A and B are $49\sqrt{3}$ and 49 N. The centre of gravity is 0·5 m from A.

Exercise 24a. (All in equilibrium.)

1. A uniform rod of mass 10 kg and length 2 m is supported at each end by a rope. If one rope is at 135° to the rod, what are the tensions in the ropes and the angle between the second rope and the horizontal rod?

2. A non-uniform rod AB of mass 0·24 kg and length 1 m is held at each end by a string. The string at A makes an angle 30° with the rod and that at B makes an angle 60°. What are the tensions if the rod is horizontal? How far is the centre of gravity from A?

3. A light bar AB has a mass of 0·1 kg suspended from it so that it hangs horizontally when held by strings at each end. The first string at A is horizontal and the second at B makes an angle 120° with the bar. What are the tensions and where is the mass hung?

4. A non-uniform bar of length 3 m and mass 12 kg is supported by ropes attached to each end, both at 30° to the bar. The bar hangs at 30° to the horizontal. What are the two tensions and how far is the centre of gravity from the lower end of the bar?

5. A light bar of length 2·5 m is held horizontally by strings at each end making angles 155° and 110° with the bar, and a mass of

2 kg. What are the tensions in the strings and how far is the mass placed from the nearest end of the bar?

6. A uniform rod of mass 0·2 kg and length 1·5 m is supported at each end by strings making angles 150° and 120° with the rod ends *A* and *B* respectively. It is kept horizontal by a vertical upward pull at 0·3 m from end *A*. Find the value of this pull and the tensions in the strings.

24.2

A particle or block is an object whose size can be neglected. Problems involving particles cannot include taking moments and so they are solved by resolving in any two directions.

A third equation is often obtained by the introduction of friction.

Let us consider a block, mass in kg of *M*, at rest on a rough plane inclined at an angle of 30° to the horizontal.

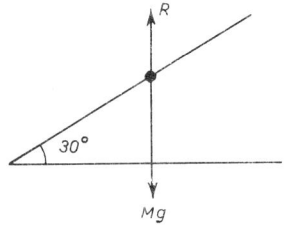

The block has a weight in N of *Mg* vertically downwards so the plane must be supporting it with a force *R* acting vertically upwards, where $R = Mg$. The force *R* can be resolved into two components, the normal reaction, *N*, which is perpendicular to the plane and the force of friction, *F*, which is acting up the plane.

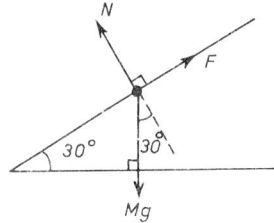

Resolving perpendicularly to the plane:

$$N = Mg \cos 30° = Mg\frac{\sqrt{3}}{2}$$

Resolving along the plane:

$$F = Mg \sin 30° = \tfrac{1}{2}Mg$$

The plane can hold the body because it is rough enough to exert this force of $\frac{1}{2}Mg$ on the body. If the plane is placed in a horizontal position the weight of the body is supported by the normal reaction.

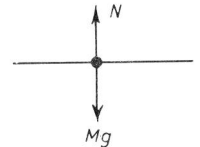

As the plane is raised at one end the body will tend to move down the plane but it is held in equilibrium by friction which adjusts itself to the required value.

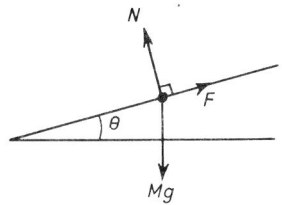

When the plane is at an angle θ the force of friction up the plane will be $F = Mg \sin \theta$.

As the plane is raised it will reach a position, at an angle α to the horizontal, such that the body is about to slip down the plane. At this point the force of friction has reached its maximum value and is said to be limiting. The angle will depend on the material and condition of the plane and body. For any given surfaces the value

$$\frac{\text{Limiting friction}}{\text{Normal reaction}}$$

is a constant μ, the coefficient of friction.

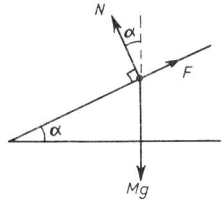

Resolving along the plane:

$$F = Mg \sin \alpha$$

Resolving perpendicularly to plane:

$$N = Mg \cos \alpha$$

Because the body is about to move the friction is limiting which gives us a third equation $\mu = \dfrac{F}{N}$, or $F = \mu N$.

Notice that if we divide the first resolved equation by the second we obtain $\dfrac{F}{N} = \dfrac{\sin \alpha}{\cos \alpha} = \tan \alpha$; showing that the angle of the plane at the point of slipping depends on the coefficient of friction.

If several forces act on the body they will tend to move it in a certain direction. Friction always acts in a direction which will oppose such a tendency to move.

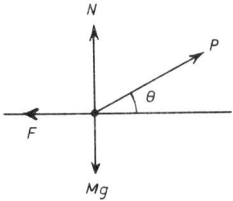

The horizontal component $P.\cos \theta$ is tending to move the body to the right and so friction acts to the left to maintain equilibrium, giving $F = P.\cos \theta$; also $N + P.\sin \theta = Mg$.

A body of mass 5 kg is resting on a plane inclined at 45° to the horizontal. What force is required to act up the plane, having a coefficient of friction of $\frac{1}{2}$, if the body is about to slip (a) down the plane, (b) up the plane?

(a) The plane has reached an angle beyond which it can support the mass by itself ($\tan 45° = 1$, which is greater than μ). An extra force in newtons of P is needed to hold the mass.

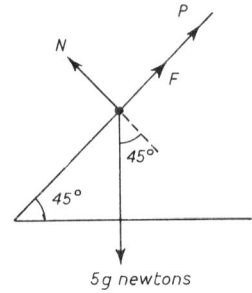

5g newtons

Resolving along the plane:

$$P + F = 5g \sin 45° = \frac{5g}{\sqrt{2}} \qquad (1)$$

Resolving perpendicularly to the plane:

$$N = 5g \cos 45° = \frac{5g}{\sqrt{2}} \qquad (2)$$

Friction is limiting:

$$\therefore F = \tfrac{1}{2}.N \qquad (3)$$

Eliminating N from (2) and (3):

$$F = \frac{5g}{2\sqrt{2}} \qquad (4)$$

Eliminating F from (3) and (4):

$$P + \frac{5g}{2\sqrt{2}} = \frac{5g}{\sqrt{2}}$$

$$\therefore P = \frac{5g}{\sqrt{2}}(1 - \tfrac{1}{2}) = \frac{5g}{2\sqrt{2}} = \frac{5\sqrt{2}}{4}g \text{ N}$$

(b) Here the force P is so great that it is about to pull the mass up the plane, and so friction acts downwards.

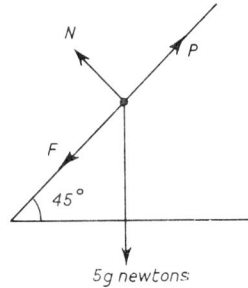

Resolving along the plane:

$$P - F = 5g \sin 45° = \frac{5g}{\sqrt{2}} \qquad (1)$$

Resolving perpendicularly to the plane:

$$N = 5g \cos 45° = \frac{5g}{\sqrt{2}} \qquad (2)$$

Friction is limiting:

$$\therefore F = \tfrac{1}{2}N \qquad (3)$$

Thus
$$F = \frac{5g}{2\sqrt{2}}$$

giving
$$P - \frac{5g}{2\sqrt{2}} = \frac{5g}{\sqrt{2}}$$

$$P = \frac{15\sqrt{2}}{4}g \text{ N}$$

Exercise 24b. (All in equilibrium, leave forces in terms of g.)

1. A particle of mass 0·5 kg is resting on a plane inclined at an angle of 24° to the horizontal. What are the normal reaction and friction? Find the coefficient of friction if the particle is about to slip down the plane when its angle with the horizontal is 37°.

2. A body of mass 9 kg is resting on a horizontal floor. The coefficient of friction is $\frac{3}{4}$. Find the value of the force acting upwards at 45° to the floor, if the mass is on the point of slipping.

3. Repeat Question 2 with the force acting downwards on to the mass at 45° to the floor.

4. A mass of 4 kg is placed on a plane inclined at 30° to the horizontal. If the coefficient of friction is $\frac{1}{5}$ find the horizontal force which is needed to hold the body when it is on the point of slipping down the plane.

5. Repeat Question 4 if (a) the body is about to move up the plane, (b) there is no force of friction needed.

6. A horizontal force of $2g$ N can just support a block of mass 3 kg which is resting on an inclined plane with coefficient of friction $\frac{1}{5}$. Calculate the angle of inclination of the plane. Find the value of the horizontal force if the block is about to move up the plane.

24.3

When we considered an extended body, such as a rod or ladder, we found that we could obtain three equations. If it rests on or against a rough surface and is about to slip we have an extra equation $F = \mu N$. In some examples the body rests against a peg, which is taken to 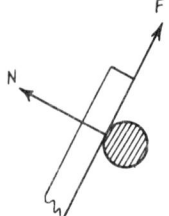 be circular. The rod forms a tangent to the peg and so the normal reaction is perpendicular to the rod and friction is along the rod. For any smooth surface the frictional force is zero, leaving a normal reaction only.

A uniform bar 0·8 m long is resting at an inclination of 30° to a rough horizontal

floor. It is supported by a smooth peg which is 0·6 m from its lower
end. Find the coefficient of friction if it is about to slip.

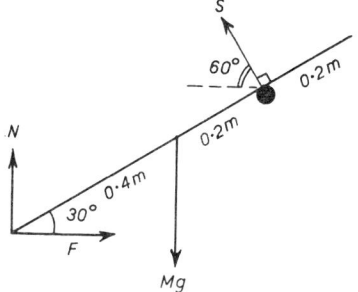

To find μ we must resolve so that N and F are in different equations.
Resolving horizontally:

$$F = S.\cos 60° = \frac{S}{2} \qquad (1)$$

Resolving vertically:

$$N = Mg - S.\sin 60° = Mg - \frac{\sqrt{3}.S}{2} \qquad (2)$$

Taking moments about the foot of the ladder:

$$S \times 0.6 = Mg \times 0.4 \cos 30° = Mg \times 0.4 \times \frac{\sqrt{3}}{2}$$

$$\therefore Mg = \sqrt{3}.S$$

Substituting in (2) for Mg:

$$N = \sqrt{3}.S - \frac{\sqrt{3}.S}{2} = \frac{\sqrt{3}.S}{2} \qquad (3)$$

Dividing (1) by (3):

$$\mu = \frac{F}{N} = \frac{\dfrac{S}{2}}{\dfrac{\sqrt{3}.S}{2}} = \frac{1}{\sqrt{3}} = \frac{\sqrt{3}}{3}$$

The coefficient of friction is $\dfrac{\sqrt{3}}{3}$.

Exercise 24c. (All in equilibrium, leave forces in terms of g.)

1. A uniform rod of mass 6 kg and length 5 m is resting horizontally against a rough vertical wall held by a rope 6 m long. One
end of the rope is joined to the rod 3 m from the wall and the

other end is joined to the wall vertically above the end of the rod. What is the coefficient of friction if the rod is about to slip?

2. A uniform ladder of mass 50 kg and length 4 m leans against a smooth wall. Its lower end rests on a rough level floor at an angle of 75°. Find the coefficient of friction at the rough surface if the ladder is about to slip.

3. Repeat Question 2 for a ladder with C. of G. 1 m from lower end.

4. A uniform beam of mass 30 kg has its lower end resting on a rough horizontal floor. The coefficient of friction is $\frac{1}{2}$. It is held in position by a horizontal force acting at the free upper end. If it is on the point of slipping find the horizontal force and the angle of inclination of the beam.

5. Repeat Question 1 with a mass of 1 kg hanging from the rod where it is joined by the rope.

6. Repeat Question 2 if a man of mass 75 kg is standing upright on a rung 1 m from the upper end of the ladder.

25. Equilibrium by Graphical Methods

25.1

Graphical methods can be used to solve some of the problems involving forces acting on a particle. In section 23.2 we found that the polygon of forces gave the resultant force. If a body is in equilibrium there will be no resultant force and so the polygon of the original forces is closed, the arrows following round.

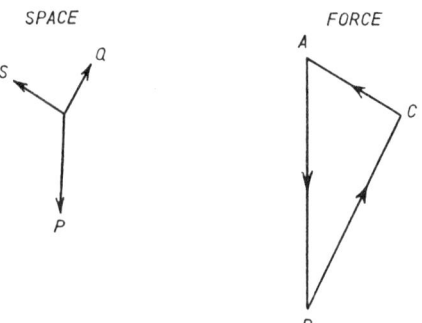

A force of 10 newtons acts on a particle. A second force of 6 N acts at 130° to the first. Find the force necessary to keep the particle in equilibrium and the angle it makes with the first force.

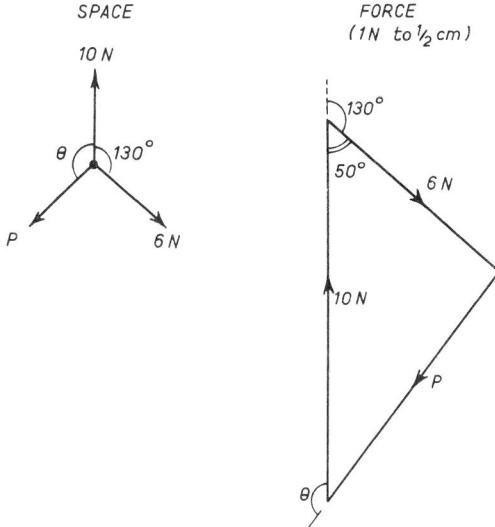

In the original (scale 1 N to 1 cm) the 10 cm line was drawn first to represent the 10 N, the 6 cm line must be drawn from the top of the first line so that the arrows go round the triangle in the same direction. The third side of the triangle is completed to represent P. Its length is 7·7 cm and θ is 143°.

A force of 7·7 N at 143° to the first force keeps the particle in equilibrium.

Some problems which involve finding a least force can be solved by semi-graphical methods in which the force diagram need not be drawn accurately.

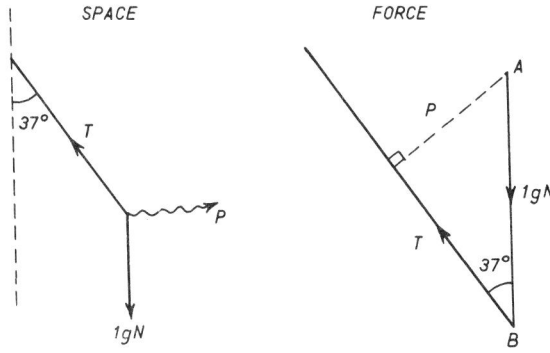

Find the least force needed to pull the bob of a pendulum of mass 1 kg, so that its string is making an angle of 37° with the downward vertical. What is the tension in the string?

Since the direction of P is not known it can be shown with a wavy line arrow. In the freehand diagram AB is fixed to represent $1g$ N. From B a line can be drawn at 37°, in the direction of the tension in the string T. The full lines in the force diagram show the reasoning up to this point. The line representing P will pass through A and will join the other line which represents T. If P is to be a minimum its line must be as short as possible, i.e. the perpendicular dotted line from A. From the triangle $P = 1g \sin 37°$, and $T = 1g \cos 37°$.

P is a force of $0 \cdot 602g$ newtons at 90° to the string which has a tension (T) of $0 \cdot 799g$ N.

Exercise 25a. (All in equilibrium.)

1. A particle is acted on by a force of 7 N and a force of 5 N with an angle 60° between them. Find the magnitude of the force which will keep the particle in equilibrium, and the angle it makes with the 7 N.

2. Repeat Question 1 with the angle 135°.

3. A particle of mass 0·5 kg is supported by a string. The particle is drawn aside by a horizontal force S until the string makes an angle of 53° with the downward vertical. Find the force S and the tension in the string.

4. Find the least value of S, if it is no longer horizontal, and its direction in the previous question, also the tension in the string.

5. Find the least value of T, and the value and the direction of S in Question 3 if the angle changes.

6. A mass of 2 kg is supported by two strings, one of which is inclined at 40° to the vertical. Find the direction and magnitude of the least force exerted in the other string.

25.2

A graphical or semi-graphical method can be used to solve problems of equilibrium of bodies if the number of forces acting on the body can be reduced to three.

Let us consider an L-shaped bracket supported at the corner by a smooth hinge. The word smooth indicates that it will turn freely. It is held in equilibrium with one arm vertical, by a horizontal force of 2 N. At the horizontal end is a vertical force P. The vertical arm is 12 cm long and the horizontal arm is 5 cm long.

Since the hinge reaction is unknown in magnitude and direction we start by taking moments about H:

$$2 \times 12 = P \times 5$$

$$4 \cdot 8 \text{ N} = P$$

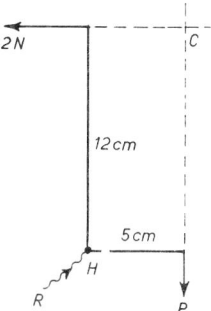

Now let us consider the point C where the lines of action of the two known forces cross. If we take moments about C we find that P and the 2 N have zero moment. The body is in equilibrium so the sum of the moments must be zero. Since we have only one other force R, the reaction of the hinge, its moment about C must also be zero for there is no other force to balance its moment. This can only happen if the line of action of R passes through C. This demonstrates the important principle that if three non-parallel forces hold a body in equilibrium their lines of action must be concurrent (i.e. pass through the same point).

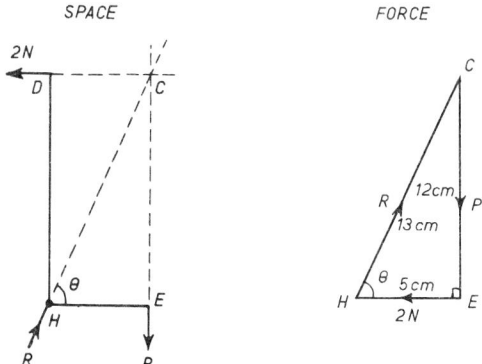

When the line of action of R has been drawn through C we see that triangle HCD or HCE can represent the force diagram since their sides are in the same directions as the three forces. By Pythagoras'

theorem $HC = 13$ cm. The distances in the force diagram represent the forces.

$$\therefore \frac{2}{5} = \frac{P}{12} = \frac{R}{13}$$

This confirms that $P = 4\cdot8$ newtons.

$$R = \frac{26}{5} = 5\cdot2 \text{ newtons}$$

and its angle θ with the horizontal is given by

$$\tan \theta = \frac{12}{5} = 2\cdot400$$

The reaction of the hinge is $5\cdot2$ N at an angle $67° \ 23'$ to the horizontal.

The force and space diagrams can be combined into one accurate scale drawing if the dimensions are not convenient numbers.

Exercise 25b. Repeat Questions 1 to 5 of Exercise 24a, using graphical or semi-graphical methods.

6. A uniform bar AB of mass 10 kg is hinged at A to a vertical wall. A string is attached to B and is fastened at the other end to point C above A on the wall, such that ABC is an equilateral triangle. Find by drawing the magnitude and direction of the hinge's reaction and the tension in the string when the bar is in equilibrium.

25.3

Problems involving friction can often be reduced to three forces by combining the friction (F) and normal reaction (N) into a single reaction (R) such that F and N are the resolved parts of R. This is especially useful when friction is limiting, as shown below.

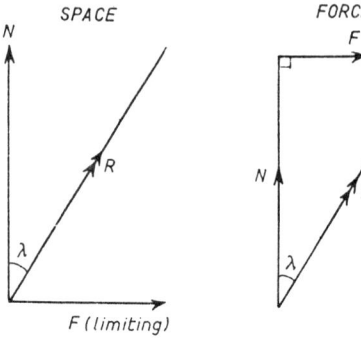

Suppose that R makes an angle λ with the normal. From the force diagram $\tan \lambda = \dfrac{F}{N} = \mu$. The angle λ is called the angle of friction.

We shall now consider the example of section 24.3 treating N and F as a composed force R. There are three forces only, so they are concurrent.

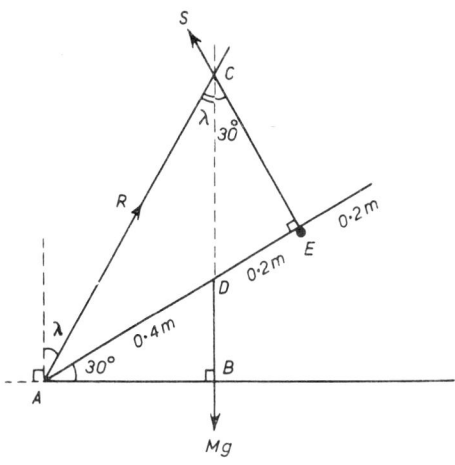

In $\triangle ABC$:

$$\mu = \tan \lambda = \frac{AB}{CB} = \frac{AB}{CD + DB}$$

But in $\triangle ADB$:

$$DB = 0{\cdot}4 \sin 30^\circ = 0{\cdot}2 \text{ m};$$
$$AB = 0{\cdot}4 \cos 30^\circ = 0{\cdot}2\sqrt{3} \text{ m}$$

In $\triangle DCE$:

$$CD = 0{\cdot}2 \operatorname{cosec} 30^\circ = 0{\cdot}4 \text{ m}$$

$$\therefore \tan \lambda = \frac{0{\cdot}2\sqrt{3}}{0{\cdot}4 + 0{\cdot}2} = \frac{0{\cdot}2\sqrt{3}}{0{\cdot}6} = \frac{\sqrt{3}}{3}$$

$$\therefore \mu = \frac{\sqrt{3}}{3}$$

Notice that if a line was drawn from D perpendicular to the rod to meet AC at F, then CDF would be one possible triangle of forces if we had been asked to find S and R in terms of Mg.

In most problems it is easier to draw an accurate diagram and find the answers by measurement.

Exercise 25c. Repeat Exercise 24c, Questions 1–3 by graphical or semi-graphical methods and Question 4 by graphical methods.

5. A mass of 5 kg is placed on a plane inclined at an angle of 45° to the horizontal. If the coefficient of friction is $\frac{3}{4}$, find the magnitude and direction of the least force needed to prevent the mass slipping down the plane.

6. Repeat Question 5 finding the magnitude and direction of the least force needed if the mass is about to move up the plane.

26. Force, Momentum, Energy, Work, Power

26.1

Newton's First Law of Motion states that every body perseveres in its state of rest or of moving uniformly in a straight line, except in so far as it is made to change that state by external forces.

Thus for a body to change its velocity, which means that it will be accelerating, an external force must be applied.

Newton's Second Law of Motion states that change of motion is proportional to the impressed force and takes place in the direction in which the force is impressed.

This law leads to the deduction that the acceleration (a) is proportional to the net force acting (F) and is in the direction of that force.

$$F \propto a$$

Mass is difficult to define. It can be called the amount of matter in a body. The first law is sometimes called the law of inertia. The measure of inertia is mass and the greater the inertia of a body the greater the force needed to change its state of rest or uniform motion. The force F therefore depends on the mass of the body (m)

$$F \propto m$$

Thus F, the force necessary to give a mass m an acceleration a, is proportional to both these quantities. In the absolute international system of units (SI)

$$F = ma$$

The acceleration produced in metres per second^2 is a when a net force in newtons of F is acting on a mass in kilogrammes of m, the constant of proportionality being unity. Both the laws and the deductions from them are based on experimental results.

A diesel passenger train of mass 60 000 kg has a motor which exerts a pull of 40 000 N. Find the acceleration of the train if the resistance to motion is 10 000 N.

The resistance will be friction and air resistance.

The train is treated as a particle.

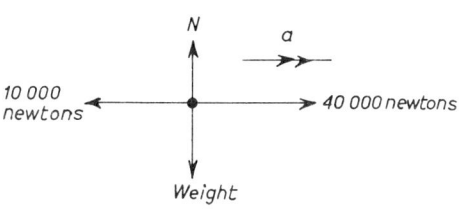

All forces are shown but we need to consider the horizontal resolution only. Net force to right is $40\,000 - 10\,000 = 30\,000$ N. The mass is 60 000 kg.

Let the acceleration be a:

$$F = ma$$

i.e. $$30\,000 = 60\,000a$$

$$a = 0\cdot5 \text{ m/s}^2$$

Exercise 26a

1. A mass of 1 kg is accelerating at $9\cdot8$ m/s^2. Find the net force acting.

2. An object moves horizontally with an acceleration of 2 m/s^2 when a force of 20 N is exerted on it. Calculate its mass.

3. A particle of mass 245 g is accelerating at 12 cm/s^2. What is the net force in newtons acting?

4. A force of 120 N is exerted on a mass of 50 g. What is its acceleration in m/s^2?

5. A car of 1500 kg has a motor which exerts a pull of 4000 N. If the resistance to motion is 1500 N, find the acceleration of the car.

6. A train of 200 000 kg has an engine which exerts a pull of 60 000 N. If the acceleration produced is $0\cdot25$ m/s^2, what is the resistance to motion in newtons?

26.2

A very special force is a body's weight. This is the force exerted vertically downwards on a body due to its attraction to the earth. In SI units a body of mass 1 kg has a weight of 9·8 N, for in free fall $a = g = 9·8$ m/s². Thus a man of mass 70 kg has a weight of $70 \times 9·8 = 686$ newtons.

Newton's Third Law of Motion states that action and reaction are equal and opposite.

Let us consider a mass in kg of m standing on a weighing machine in a lift which is at rest.

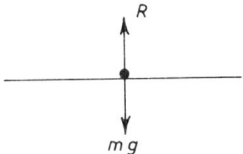

It has a weight in newtons of mg acting vertically downwards, and since it is in equilibrium the machine exerts a reaction, R, which is equal and opposite to mg. The diagram is drawn showing the forces acting on the mass. It is only by Newton's Third Law that we can say the mass exerts a force R on the machine.

At an instant when the lift is moving with an acceleration, in m/s², of $2f$ the weighing machine registers $20g$ N. At a later instant when it is moving down with an acceleration of f it registers $11g$ N. Find m and f.

The diagram shows the forces acting on the mass when moving upwards, being its weight downwards and the weighing machine reaction upwards.

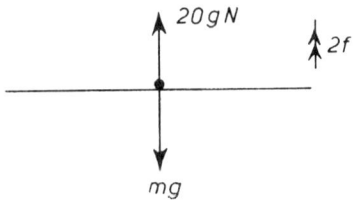

The net upward force $F = 20g - mg$

since $F = \text{mass} \times \text{acceleration}$

$$20g - mg = m2f \tag{1}$$

The next diagram shows the forces acting at the later instant when the mass is moving downwards.

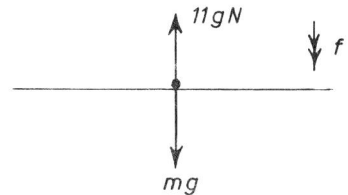

The net downward force $= mg - 11g$

$$\therefore mg - 11g = mf \qquad (2)$$

Doubling (2) $\qquad\qquad 2mg - 22g = 2mf$

Subtracting (1) $\qquad\quad 3mg - 42g = 0$

$$\therefore m = 14 \text{ kg}$$

Substituting in (2) $\qquad 14g - 11g = 14f$

$$3g = 14f$$

$$f = \frac{3 \times 9 \cdot 8}{14} = 2 \cdot 1 \text{ m/s}^2$$

Exercise 26b

1. A boy of mass 50 kg is standing on a weighing machine in a lift. Find the reading of the machine in newtons when:

   (a) the lift is stationary;
   (b) it ascends with acceleration of $1 \cdot 2$ m/s^2;
   (c) it moves with uniform velocity;
   (d) it ascends with retardation of $1 \cdot 2$ m/s^2.

2. A stone of mass $0 \cdot 098$ kg falls freely through the air with an acceleration $9 \cdot 6$ m/s^2. What is the air resistance?

3. A stone of mass 4 kg falls in water against a resistance of 5 N. What is its acceleration?

4. A body is hung from a spring balance in a lift. When the lift is at rest the reading is $4 \cdot 9$ N. If the reading is 4 N, when the lift is moving, find the acceleration and explain why the lift can be moving upwards or downwards.

5. A rain drop of mass 100 mg is falling with constant velocity, calculate the upward air resistance in newtons. Find the value if the drop accelerates at 98 mm/s^2: (a) downwards, (b) upwards.

6. When a lift descends with an acceleration f a mass m exerts a force on the lift of 100 N. When the lift ascends with a retardation of $2f$ the force exerted is 80 N. Find m and f.

26.3

We shall now consider a body with mass in kg of m moving with a velocity in m/s of v.

The momentum is defined as the product of the mass and the velocity.

$$\text{Momentum in newton seconds} = mv$$

The impulse in Ns of a constant force, in newtons, F acting for a time, in seconds, t is defined as Ft.

But from section 26.1

$$F = ma$$

$$\therefore \text{Impulse} = Ft = mat$$

But the force is constant, therefore the acceleration is constant. By section 22.3

$$at = (v-u)$$

$$\therefore \text{Impulse} = m(v-u) = mv-mu$$

The impulse of a force = the change of momentum produced by it.

A cricket ball of mass 0·15 kg is moving horizontally at 38 m/s and is hit back horizontally at 22 m/s. If contact between ball and bat lasted for 0·02 of a second, find the average force exerted by the bat.

$$\text{Mass} = 0·15 \text{ kg}$$

Considering the direction of the bat's force:

Initial velocity:

$$u = -38 \text{ m/s}$$

Final velocity:

$$v = +22 \text{ m/s}$$

$$\therefore Ft = m(v-u)$$
$$F \times 0·02 = 0·15(22-(-38))$$
$$F \times 0·02 = 9$$
$$F = \frac{9}{0·02} = \frac{900}{2} = 450 \text{ N}$$

An explosion, such as takes place in a gun, gives a forward momentum to the shell equal to the backward momentum of the gun.

Exercise 26c

1. What is the momentum of a mass of 490 g moving with a velocity of 4 m/s?

2. What is the impulse that has acted to move a body from rest to 15 m/s if the body has mass 3 kg?

3. A gun of mass 900 kg fires a shell of 2 kg. If the gun moves back with an initial velocity of 6 m/s, find the initial velocity of the bullet and the constant force necessary to stop the gun in 1 second.

4. A cricket ball of mass 0·15 kg is moving horizontally with a speed of 15 m/s and is hit back horizontally with a force of 400 N. What is its return velocity in m/s if the ball and bat were in contact for $\frac{1}{50}$ sec.?

5. What is the net uniform forward force in N if a car of mass 1500 kg moves from rest to (a) 30 m/s in 8 s, (b) 135 km/h in 15 s?

6. A particle reaches a speed of 30 m/s in 20 seconds, when a force of 10 N acts. Find the mass of the particle.

26.4

A mass in kg of m moving with velocity in m/s of v has gained energy of motion, kinetic energy.

The kinetic energy $= \frac{1}{2}mv^2$ and is measured in joules, where 1 joule = 1 newton metre.

A constant force in newtons of F having moved its point of application forward through a distance in m of s along its line of action has done work in joules of Fs. If the point moves backwards work has been done against the force.

But from 26.1 $F = ma$

Therefore work done $Fs = mas$

But the force is constant, therefore the acceleration is constant. Then, by section 22.3

$$as = \frac{v^2 - u^2}{2}$$

$$\therefore \text{Work done} = \frac{m}{2}(v^2 - u^2) = \tfrac{1}{2}mv^2 - \tfrac{1}{2}mu^2$$

The work done by a force = the change in kinetic energy.

A cricket ball of mass 0·15 kg is moving with a velocity of 22 m/s.

What is its kinetic energy and the average stopping force if the hands move back 20 cm, in the direction the ball moves, during the catch?

$$\text{K.E.} = \tfrac{1}{2} \times 0\cdot 15 \times 22 \times 22 = 36\cdot 3 \text{ joules}$$

During contact with the ball the hands move 0·2 m

$$F \times 0\cdot 2 = 36\cdot 3$$

$$F = 180 \text{ N (correct to 2 s.f.)}$$

Exercise 26d

1. What is the kinetic energy of a mass of 490 g moving with a velocity of 4 m/s?

2. What is the work done in moving a body from rest to 15 m/s if the body has mass 3 kg?

3. What are the initial kinetic energies of gun and shell in Question 3 of Exercise 26c? Find the constant force needed to stop the shell in 0·6 m.

4. Find the gain of energy of the ball in Question 4 of Exercise 26c after it has been hit.

5. Find the braking force (including friction) if a car of 1500 kg is travelling at 54 km/h in neutral (i.e. without the engine pulling) and stops in 22 m.

6. A particle reaches a speed of 30 m/s in 300 m, when the force is 10 N. Find the mass of the particle.

26.5

Power is defined as the rate of doing work. A constant force in newtons of F acts for a time in seconds of t, and its point of application moves a distance in metres of s.

$$\text{Power} = \frac{\text{work done}}{\text{time taken}} = \frac{F \cdot s}{t}$$

But from section 22.3

$$\frac{s}{t} = \frac{v + u}{2}$$

$$\therefore \text{ Power in watts} = F\frac{(v + u)}{2}$$

1 watt is 1 joule per second. In our problems we consider power at an instant. We can therefore take the initial and final velocities to be the same. A force F acts at an instant when the velocity is v.

$$\text{Power in watts} = Fv$$

For larger outputs of work we will use the kilowatt (kW).

An engine works at 195 kW when it is drawing a train (including the engine) of 200 000 kg at a uniform velocity of 54 km/h along a level track. Find the resistance to motion in N per 1000 kg.

$$\text{Power} = 195 \text{ kW} = 195\,000 \text{ W}$$

Since the train is moving along a level track the only force is the resistance in N of F due mainly to friction. The velocity v is 54 km/h = 15 m/s.

Work done per second in watts $= Fv = 15F$

$$\therefore 15F = 195\,000$$

$$F = 13\,000 \text{ N}$$

The resistance is required in N for every tonne of the train, which has mass 200 tonnes; for 1 tonne = 1000 kg.

$$\therefore \text{ The resistance per tonne } = 65 \text{ N per tonne.}$$

Calculate the power needed to draw the train up an incline of 1 in 400 at the same speed if the resistance is the same.

Since the resistance is the same it will require the original 195 kW to overcome resistance. Going up an incline we have introduced another force, the weight of the train.

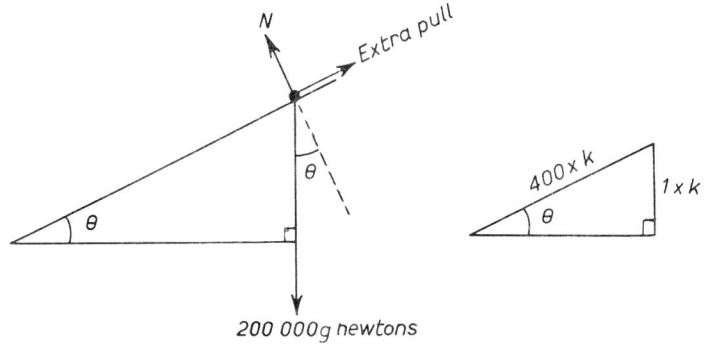

200 000g newtons

The component of weight acting down the plane is

$$200\,000g \times \sin \theta \text{ N}$$

The railways have always measured incline as the rise 1 unit in x units along the track.

$$\therefore \sin \theta = \frac{1 . k}{400 . k} = \frac{1}{400}$$

∴ Force down the plane

$$= 200\,000 \times 9{\cdot}8 \times \frac{1}{400} = 4900 \text{ N}$$

∴ Extra pull $= 4900$ N;

$$\text{Extra power} = \text{force} \times \text{velocity}$$

$$= 4900 \times 15 \text{ W}$$

$$= 4{\cdot}9 \times 15 \text{ kW}$$

$$= 73{\cdot}5 \text{ kW}$$

$$\text{Total power} = 268{\cdot}5 \text{ kW}$$

Exercise 26e

1. A car of mass 1500 kg is travelling at a steady speed of 72 km/h on the level against a resistance of 180 N per 1000 kg. Find the power at which the engine is working. Calculate the power required to drive the car up a gradient of 1 in 20 at the same speed with the same resistance.

2. A train of mass 200 000 kg travels at a steady speed of 144 km/h on the level against frictional resistance of 54 N per 1000 kg. Find the power.

   If this is the maximum power find the speed (in km/h) at which the train can travel up an incline of 1 in 50 if the resistance is unchanged.

3. A diesel train of mass 300 000 kg and motor 900 kW travels against a resistance of $\dfrac{v^2}{160}$ in N per 1000 kg when the train is moving at v in km/h. What is the maxiᵐum speed (in km/h) of the train on the level?

4. The motor of a diesel train of mass 150 000 kg can exert a pull of 50 000 N, the constant resistance to motion being 50 N per 1000 kg. Calculate the acceleration of the train (in m/s²) and the motor's power 10 seconds after starting.

5. A train of mass 200 000 kg is travelling along a level track at 90 km/h. If the resistance to motion is 60 N per 1000 kg calculate the power of the engine. Find the power required to travel up an incline of 1 in 300 at the same speed against the same resistance.

6. A car of mass 1500 kg has a maximum speed of 216 km/h, developing 216 kW. Find the pull against tractive resistance.

   Calculate the maximum gradient that the car can travel up at 54 km/h whilst still in top gear, against tractive resistance

608 N per 1000 kg if the total pull remains the same. Find the power developed.

26.6

By Newton's Third Law if two bodies collide any momentum communicated to one of the bodies is equal and opposite to that communicated to the other body. It follows that, if no external forces act, the momentum before the collision equals the momentum after the collision for bodies travelling in a straight line. We know that the kinetic energy changes, for some of it is converted into vibrations which we hear as noise.

Consider a body of mass m_1 with initial velocity before collision of u_1, colliding with a body of mass m_2 having velocity u_2. They are moving in a straight line and have final velocities v_1 and v_2 respectively after the collision.

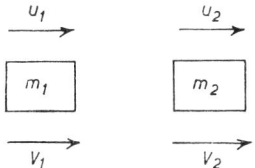

$$\text{Momentum before the collision} = m_1.u_1 + m_2.u_2$$

$$\text{Momentum after collision} = m_1.v_1 + m_2.v_2$$

$$\therefore m_1.u_1 + m_2u_2 = m_1.v_1 + m_2.v_2$$

This is called the principle of conservation of linear momentum.

A particle of mass in kg of $2m$ travelling at 6 m/s collides with a particle of mass m travelling at 8 m/s in the opposite direction. After impact the first particle travels at 1 m/s in its original direction. Find the velocity of the second particle after collision and the fraction of the kinetic energy of the system changed into other forms of energy.

Taking the velocity of the first particle as positive the second starts with a negative velocity in that direction.

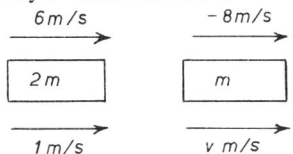

$$\text{Momentum before collision in Ns} = 2m \times 6 + m(-8) = 4m$$

$$\text{Momentum after collision in Ns} = 2m \times 1 + m.v$$

$$4m = 2m + m.v$$

$$v = 2 \text{ m/s}$$

Notice that if the bodies move in the same direction after impact the front body moves at a velocity greater than or equal to that of the back; and that it is the momentum, not the velocity, that controls the direction of movement after impact.

$$\text{K.E. in J before} = \frac{2m.6^2}{2} + \frac{m(-8)^2}{2} = \frac{m}{2}(72+64) = \frac{136m}{2}$$

$$\text{K.E. in J after} = \frac{2m.1^2}{2} + \frac{m.2^2}{2} = \frac{m}{2}(2+4) = \frac{6m}{2}$$

$$\text{K.E. in J lost} = \frac{130m}{2}$$

$$\text{Fractional loss} = \frac{130}{136} = \frac{65}{68}$$

The conditions after collision are fixed by the elasticity of the bodies, but as this is beyond the scope of our work we have to be given one of the resulting velocities, or that the bodies join together giving them a common velocity.

26.7

When a mass in kg of m falls freely it gains kinetic energy. Suppose it falls a distance in m of s in changing from an initial velocity u to a final velocity in m/s of v, whilst falling with constant acceleration in m/s^2 of g.

$$\text{The gain in kinetic energy in joules} = \frac{m(v^2-u^2)}{2}$$

The work is done by a force in N of mg whose point of application moves a distance in m of s. The body is said to have lost potential energy.

$$\text{The loss in P.E. in joules} = mgs$$

Equating $$\frac{m(v^2-u^2)}{2} = mgs$$

i.e. $$v^2 - u^2 = 2gs$$

which is one of the equations of 22.3 when acceleration $a = g$.

Therefore K.E. gained = P.E. lost and vice versa. This means that the amount of mechanical energy has remained constant, which is part of the principle of conservation of energy.

A body of mass m is thrown vertically upwards in air with a velocity of 84 m/s. How far does it travel before coming to rest for an instant?

$$u = 84 \text{ m/s} \quad v = 0$$

$$\text{K.E. lost} = \frac{m(84^2 - 0)}{2} = m \times 84 \times 42$$

$$\text{P.E. gained} = mgs = m \times 9.8 \times s$$

By the conservation of energy $\not{m} \times \overset{0.1}{\cancel{9.8}} \times s = \not{m} \times \overset{12}{\cancel{84}} \times \overset{3}{\cancel{42}}$

$$0.1s = 36 \quad \text{i.e. } s = 360 \text{ m}$$

Notice that the problems of section 22.4, not involving time, can be solved by energy considerations.

26.8

We shall now consider the motion of two connected bodies.

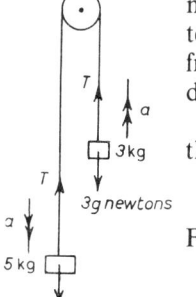

Masses of 5 kg and 3 kg are connected by a light nylon string passing over a smooth pulley. Find the tension in the string when the system is released from rest, and the velocity after the 5 kg has descended 4.9 m.

Let the tension in the string in N be T and let the acceleration in m/s² be a.

$$\text{Net force on a body} = ma$$

For the 3 kg mass:

$$\text{Net upward force} = T - 3g$$

$$\therefore T - 3g = 3a \tag{1}$$

For the 5 kg mass:

$$\text{Net downward force} = 5g - T$$

$$\therefore 5g - T = 5a \tag{2}$$

Adding (1) and (2): $2g = 8a$

$$\therefore a = \frac{g}{4} = \frac{9.8}{4} = 2.45 \text{ m/s}^2$$

Substituting in (1) $T - 3g = \frac{3g}{4}$

$$\therefore T = \frac{15g}{4} = 36.75 \text{ N}$$

For either mass:

$$a = 2{\cdot}45 \text{ m/s}^2; \quad u = 0; \quad s = 4{\cdot}9 \text{ m}$$

$$v^2 = 0 + 2 \times 2{\cdot}45 \times 4{\cdot}9 \quad \therefore \text{ Velocity is } 4{\cdot}9 \text{ m/s}$$

Alternatively considering the whole system:

$$\text{K.E. gained} = \frac{(3+5)}{2}v^2 = 4v^2$$

$$\text{P.E. lost} = 5g \times 4{\cdot}9$$

$$\text{P.E. gained} = 3g \times 4{\cdot}9$$

But K.E. gained + P.E. gained = P.E. lost

$$4v^2 + 3g \times 4{\cdot}9 = 5g \times 4{\cdot}9$$

$$v^2 = \frac{2 \times 9{\cdot}8 \times 4{\cdot}9}{4}; \quad v = 4{\cdot}9 \text{ m/s}$$

If the time was required we should have to use one of the equations of 22.3.

Exercise 26f

1. A truck of mass 10 000 kg moving at 5 m/s collides with a truck of mass 5000 kg, which is stationary. If the trucks couple together, find their velocity and the fraction of kinetic energy lost in the collision.

2. A truck of mass 6000 kg moving at 4 m/s collides with a truck of 2000 kg moving in the opposite direction at 6 m/s. If they couple together, find their velocity. Write down an expression for their kinetic energy after impact and calculate the distance required to stop them if a retarding force of 2500 N is applied.

3. A truck of 6000 kg moving at 4 m/s collides with a truck of 15 000 kg moving in the same direction at 1 m/s. If they join together, find their common velocity and the time taken to stop them if a retarding force of 10 000 N is applied.

4. A truck of mass 8000 kg moving at 6 m/s collides with a truck of mass 12000 kg moving at 2 m/s. Calculate their common velocity after impact and the fractional loss of kinetic energy if they were moving (a) in the same direction, (b) in opposite directions.

5. A mass of 0·15 kg moving with velocity 0·4 m/s collides with a mass of 0·10 kg moving with velocity 0·2 m/s. The first mass moves in its initial direction with speed 0·15 m/s. Find the velocity of the other mass if they were moving in (a) the same direction, (b) opposite directions.

6. A body is projected vertically upwards with a velocity of 42 m/s. Use energy considerations to find to what height it rises (*a*) before the velocity is halved, (*b*) before it comes to rest. Show that its speed is again 42 m/s when it returns to the starting point.

7. Masses of 4 kg and 10 kg are connected by a light nylon string passing over a smooth pulley. Find the tension in the string, also the velocity attained and time taken for the 4 kg mass to move upwards 2·45 m.

8. A mass of 4 kg rests on a smooth horizontal table and is attached by a light string, which passes over a smooth pulley at the edge of the table, to a mass of 2 kg hanging freely. If the system is allowed to move, find the velocity after the system has moved 1 m.

9. Repeat Question 8, using a rough table, coefficient of friction being $\frac{1}{4}$.

---

27. Illustration of Statistical Data

27.1

When data is presented we shall be required to derive statistics which will make clear some particular aspect of the original data. An illustration of the statistics should then make comparison of the data easier, but it must be a fair unbiased picture of the facts.

We start with the pictogram. This consists of a number of small drawings of equal size, each of which will represent a given number of the original data.

Let us consider the number of medium-size tins of soup sold by a shop in a week. Suppose the figures are Chicken 94, Mushroom 37, Tomato 50, Onion 11, and Vegetable 49. It would confuse the issue to draw 94 tins for chicken and so we let each pictogram tin represent 10 tins, and draw the final four as part of a tin.

It would be ambiguous to represent the sale as single cans of different size for each type, as it is impossible to judge the scale used; it may be by length, area, volume or in between. Thus if we have two objects, one represented by a cube of side 1 cm and the other a cube of 2 cm, without further information it may be illustrating the second as twice, four, or eight times as large as the first, or any multiple in between. If colour is used it should be the same for every pictogram, as particular colours could cause bias. Each little can drawn is called an isotype.

PICTOGRAM OF SOUP SALES

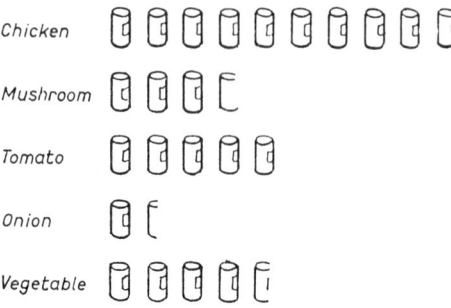

27.2

A bar chart will represent the sales as oblongs of equal width, their lengths representing the sales. If we take 1 cm to represent 10 tins the length of the bars will be 9·4, 3·7, 5·0, 1·1, and 4·9 cm respectively. Since the widths are constant, the area of the bars also represent the sales.

BAR CHART OF SOUP SALES

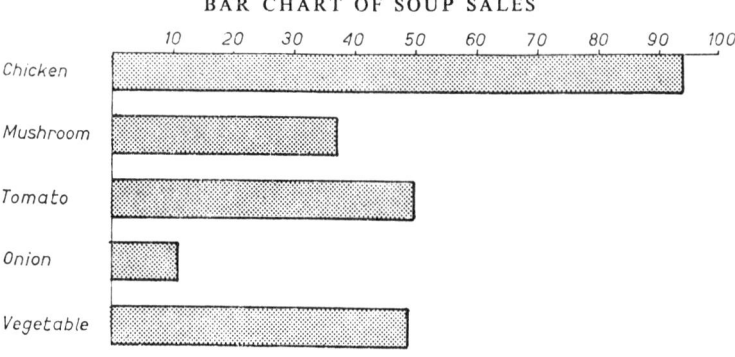

The bar chart can be used for sub-groups. Tins of soup can be of the ordinary or condensed strength and so we may be provided with further statistics:

Type	Chicken	Mushroom	Tomato	Onion	Vegetable
Condensed	30	20	25	10	19
Ordinary	64	17	25	1	30
Total	94	37	50	11	49

These can be represented on a bar chart of the same type but distinguishing the strength of each type of soup.

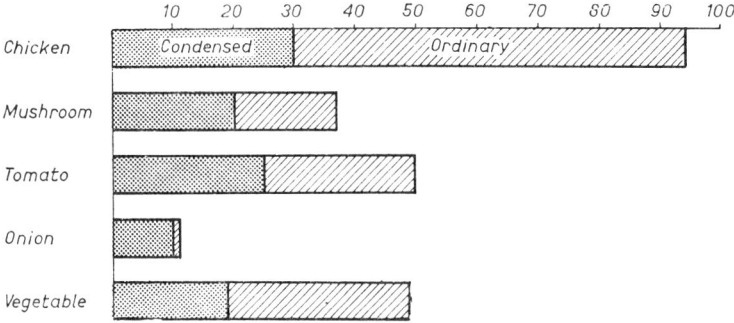

We must be careful now, the statistics have not proved that chicken is the most popular soup. Firstly, there is the special information that should have been established during the experiment. Was there a special sales week when one tin of chicken and one other tin were offered for the price of a single tin? Have all the types been on the market long enough for people to know about them? Did the shop run out of onion soup? Are there any large single orders? Having made sure that there is no special condition applying to any of the types, we can say that for this shop in the week of the experiment chicken was the most popular type.

Exercise 27a

1. In a year the Government Budget required £5325 million. Each £ of revenue was collected in the following way : taxes on income and capital $52\frac{1}{2}$p; taxes on spending $42\frac{1}{2}$p; and other revenue 5p. Construct a pictogram with a 2p coin as the basic symbol.

   Each £ collected was spent in the following way : Surplus 2p ; National Debt 12p; Defence $28\frac{1}{2}$p; Social Services 41p; and other services $16\frac{1}{2}$p. Construct a pictogram with a 2p coin as the basic symbol.

2. The following table is taken from an urban primary school roll for 1969–70.

Year	1	2	3	4	5	6
Girls	38	18	20	21	18	19
Boys	28	15	17	17	18	13
Total	66	33	37	38	36	32

   Construct a bar chart for each year showing the totals, but distinguishing between girls and boys.

3. The following list gives the users of steel in the first nine months of a year.

Hollow-ware and metal furniture .	756 000 tonnes
Motors, cycles and aircraft . .	1 297 000 tonnes
Electricity	364 000 tonnes
Building	999 000 tonnes
Shipbuilding	485 000 tonnes
Coal mining	405 000 tonnes
General engineering . . .	1 018 000 tonnes
Railways	449 000 tonnes

Construct a bar chart.

4. The following are the sales of newspapers for one week, by one shop.

The Times	240	*Daily Mail*	3720
Daily Telegraph	990	*Daily Sketch*	780
Daily Express	4200	*Daily Mirror*	4590
The Sun	2040	*The Guardian*	120

Represent these as a pictogram.

5. The following table shows the distribution of children's homes for children attending a county school on the Southern border of the county.

	Boys	*Girls*	*Total*
Town	54	67	121
County (North)	35	34	69
County (West)	9	6	15
County (East)	28	21	49
Total	126	128	254

Represent this as a bar diagram, each bar showing the boys, girls, and the total.

6. Construct a pictogram of the school roll for 1969–70.

Form:	6th	5th	4th	3rd	2nd	1st
Pupils:	22	38	34	40	64	53

27.3

Let us take the soup sales and compare them with the same sales a year previously.

	Chicken	*Mushroom*	*Tomato*	*Onion*	*Vegetable*	*Total*
1st year	42	33	49	15	41	180
2nd year	94	37	50	11	49	241

Each year will be represented by a pie-chart. The total represents a whole circular pie, angle 360° at the centre. Each type is represented by a slice of pie. For the first year 180 tins is represented by 360°, therefore the 42 chicken tins are represented by 84°. In practice the angle will have to be worked out by logarithms and rounded off to the nearest degree. The table then becomes

	Chicken	Mushroom	Tomato	Onion	Vegetable	Total
1st year	84°	66°	98°	30°	82°	360°
2nd year	140°	55°	75°	17°	73°	360°

For either year the angles at the centre, and therefore the areas of the segments, represent the sales. If we are to be able to compare from year to year the areas of the two circles must be in the ratio of the total sales. Let the first-year circle have radius r_1 and the second-year circle have radius r_2. Then

$$\frac{\pi r_1^2}{\pi r_2^2} = \frac{180^3}{241^4} \quad \therefore \frac{r_1}{r_2} = \frac{\sqrt{3}}{2} = \frac{1\cdot73}{2}$$

If $r_1 = 3\cdot5$ cm, then $r_2 = 4\cdot0$ cm for the original diagram.

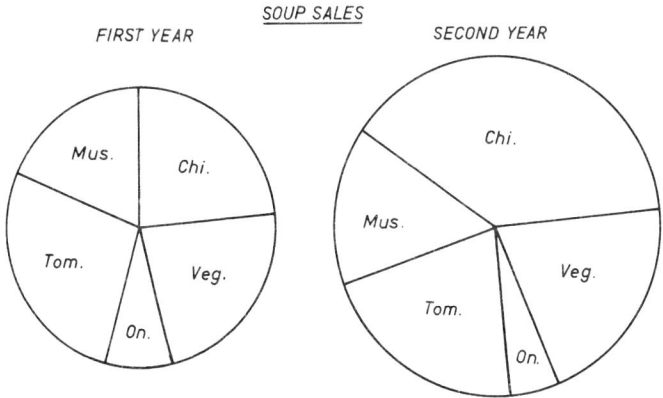

SOUP SALES

FIRST YEAR   SECOND YEAR

In these pie diagrams any constituents can be compared by comparing the areas. They bring out both the increases in individual items and also the relative sizes of the items in each week.

27.4

Very often data is given for a period of time, and statistics have to be derived to bring out some aspects of the change. It may be the proportion of one item in relationship to the total, showing how it varies from year to year, week to week, or day to day.

Let us consider the weekly sales of soup over a ten-week period.

Week:	1	2	3	4	5	6	7	8	9	10
Chicken soup:	94	112	104	140	40	78	115	212	260	112
Total soup:	241	256	260	307	160	212	274	350	403	262
Percentage chicken:	39	44	40	46	25	37	42	61	65	43

The last line shows the percentage of the total which was chicken soup. We now plot a graph of the percentage against the time. Notice that the scales must be evenly spaced. Very often graphs appear with uneven time intervals which gives a biased picture of the trend. It is best to start the percentages from zero as this does not exaggerate the changes.

CHICKEN AS PERCENTAGE OF TOTAL SALES

Using the same data it may be necessary to bring out the change in the total number of items sold from week to week. In this case any week's total would be expressed as a percentage of the previous week's total. The percentage increase can then be shown, decreases being negative. A graph would then be plotted of the percentage change against time.

Week:	1	2	3	4	5	6	7	8	9	10
Total soup:	241	256	260	307	160	212	274	350	403	262
Percentage of previous:		106	$101\frac{1}{2}$	118	52	$132\frac{1}{2}$	129	128	115	65
Percentage change:		6	2	18	-48	33	29	28	15	-35

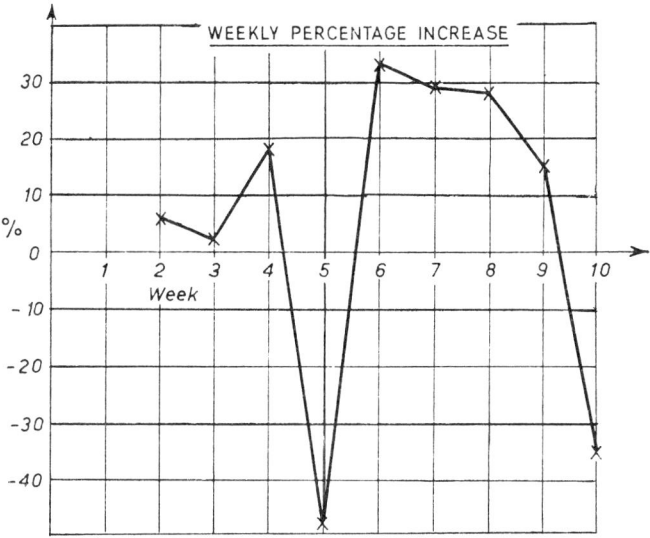

27.5

You will be able to find many examples of biased illustration of statistics. Let us consider one. The following is taken from an advertisement for dairy food and is drawn to half the scale of the original for both axes.

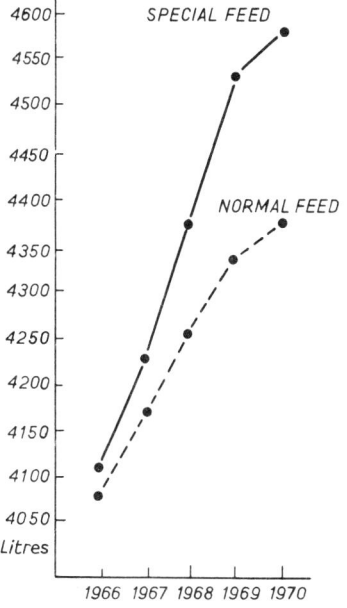

This graph reveals the remarkable increase in yields of the five chief breeds that has taken place over the last five years. Note how each year and every year the yields of those having special feed tower above the rest. Isn't this sufficient evidence to warrant you following the example of so many progressive dairy farmers and changing to the special dairy food?

The graph does have a scale on the y-axis, and the time scale is even, which is more than many graphs show. Let us consider the statement that the special feed graph towers above the other. It does as drawn, but notice that the scale starts at 4050. The 1970 figures being compared show that the ordinate of the special feed is half as long again as the ordinary feed. A towering 50% increase in yield? Not at all, the increase is about 200 in 4380, i.e. $4\frac{1}{2}$%. If drawn with the yield figures starting at zero the true comparison could be made. Hardly sufficient evidence. Notice also the claim that it is the progressive farmers who have changed to the special feed.

The basic data concerning soup sales have been used to derive statistics to bring out various aspects of the data, and these have been illustrated in various ways. When this has been done they may support certain probable conclusions, but beware of the implied statement that statistics prove this or that.

Exercise 27b

1. The following are the distributions of pupils in September of 1965 and 1970.

Form:	1	2	3	4	5	6	Total
1965:	44	37	31	34	25	23	194
1970:	53	64	40	34	38	22	251

   Construct suitable pie-charts to illustrate this data.

2. The following show the number of girls and boys in a school on September 30th for the last 10 years.

Year:	1961	1962	1963	1964	1965	1966	1967	1968	1969	1970
Girls:	89	100	93	103	118	121	117	121	134	141
Boys:	101	95	94	76	76	91	90	92	108	110

   Draw a graph showing the variation in the proportion of pupils who are girls during the period 1961–70.

3. Use the data of Question 2 to draw a graph showing the percentage increase in boys from year to year from 1962–70.

4. The following table gives the house points obtained by the three school houses over the last three years:

	1968	1969	1970
Persians	476	527	436
Romans	392	550	457
Grecians	384	472	470
Total	1252	1549	1363

Represent these figures diagrammatically to show the change in distribution of points.

5. Using the data of Question 5, Exercise 27a, draw a pie-chart to show the distribution of boys in the four districts. Draw another pie-chart of the correct size to show the distribution of girls.

6. The following is a salary extract.

Year ending April:	1965	1966	1967	1968	1969	1970
Salary:	£461	£752	£812	£920	£1023	£1119
Income tax:	£9	£31	£41	£38	£59	£75

Illustrate (i) the proportion of income taken in tax, (ii) the percentage increase in (*a*) salary, (*b*) tax from year to year. (Use the same graph for (*a*) and (*b*).)

28. Mode, Median, Quartiles

28.1

A box of matches often states the average contents on the label. Let this be 40 sticks. Suppose that we counted the sticks in each of 200 boxes. The results would be gathered in a frequency table. The first row shows the number of sticks in a box and the second shows the

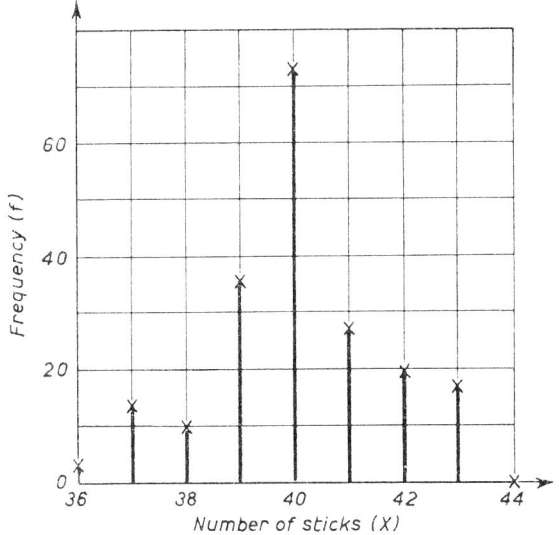

number of boxes with that number of sticks in each, this row is called the frequency and adds up to 200, the total number of boxes.

Number of sticks (X):	36	37	38	39	40	41	42	43	44
Number of boxes (f):	3	14	10	36	73	27	20	17	0

The variation of sticks is discrete, which means that the number of sticks is exact and there is no measure between any one and the next. Thus the number of sticks cannot fall between 36 and 37, etc. As this is a discrete variation we plot a frequency diagram.

Here we can introduce one of the statistical averages – the mode. The mode is the most popular class or group, i.e. 40 sticks per box. There are three boxes containing 36 sticks but none with 44. This may be because it is impossible to get 44 sticks into a box.

28.2

We may now derive a table which adds each frequency to the sum of the previous frequencies, and plot the results as a cumulative frequency diagram.

	Up to							
Number of sticks:	36	37	38	39	40	41	42	43
Cumulative frequency:	3	17	27	63	136	163	183	200

Thus there are 136 boxes which contain up to 40 matches, but 63 that contain less than 40 (i.e. up to 39).

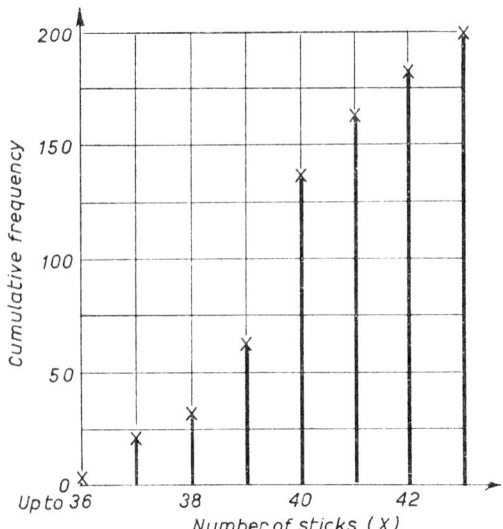

Exercise 28a. Plot a frequency diagram and cumulative frequency diagram for the following tables of data:

1. The number of beans in each of 100 packets was counted with the following results:

No. of beans:	24	25	26	27	28	29	30	31	32	33
No. of packets:	6	8	9	12	18	21	12	8	0	6

2. The numbers of times that each of 200 children was late for school in one term are shown below:

No. of times late:	0	1	2	3	4	5	6	7	8	9	10
No. of children:	21	47	53	49	16	4	5	3	1	0	1

What is the mode?

3. The numbers of times that each of 303 children stayed to school lunches in a period of two weeks (10 school days) are listed below.

No. of meals:	0	1	2	3	4	5	6	7	8	9	10
No. of children:	68	8	5	0	23	50	8	10	32	42	57

28.3

If we are given a set of figures, for example the mathematics percentages obtained in an examination by 17 boys, they will probably be given for the alphabetical order of the boys' names. Suppose they are 64, 41, 54, 62, 81, 38, 44, 56, 32, 46, 50, 79, 65, 53, 57, 67, and 40. They must be arranged in a rank order, in this case an order of merit:

81, 79, 67, 65, *64*, 62, 57, 56, **54**, 53, 50, 46, *44*, 41, 40, 38, **32**

In this array the upper limit is **81**, the lower limit **32**, and the range is $81 - 32 = 49$.

The median is defined as the value of the middle number. The median is *54*, as this is nine numbers from each end of the 17 numbers, counting the limits as the first numbers. It is the $\dfrac{(17+1)}{2}$ th number from either end. In general, if there are N numbers the median is the value of the $\dfrac{(N+1)}{2}$ th number. If N is even, this gives a half number, and the median is taken as the mean of the two middle numbers, i.e. for 18 numbers the median is half the value of the ninth plus the tenth numbers.

The upper quartile, q_2, is the value of the mark in the middle of the upper half range from the upper limit to the median. The upper quartile is *64*, as this is five numbers from the upper limit and the median. For our 17 numbers the upper quartile is the first and the median the $\frac{(17+1)}{2}$th number. The middle of these must be

$$\tfrac{1}{2}\left(\frac{17+1}{2}+1\right) \;=\; \tfrac{1}{2}\left(\frac{17+3}{2}\right)$$

i.e. the fifth number from the upper end. In general, if there are N numbers the upper quartile is the value of the $\frac{(N+3)}{4}$th number from the upper end. This can result in quarters, halves or three-quarters as part of the number. We can deal with the halves as we did for the median. Let us now consider 18 numbers. The upper quartile is the $\frac{18+3}{2}$th, i.e. the $5\tfrac{1}{4}$th number from the upper end. It would be a quarter of the way from the fifth to the sixth number. The answer must then be rounded off to the accuracy of the original numbers, which in this section is the nearest whole number. The upper quartile is also the $\frac{(3N+1)}{4}$th number from the lower end.

The lower quartile, q_1, is the value of the mark in the middle of the lower half range from the median to the lower limit. The lower quartile is *44*, as this is five numbers from the median and the lower limit. Following the reasoning of the previous paragraph, for N numbers the lower quartile is the value of the $\frac{(N+3)}{4}$th number from the lower end.

The quartile deviation or semi-interquartile range is

$$\tfrac{1}{2}(q_2-q_1) \;=\; \tfrac{1}{2}(64-44) \;=\; 10$$

This gives a measure of the spread of marks about the median; a quarter of the marks lie within 10 marks above the median and another quarter within 10 below the median. If q_2 and q_1 are not exact numbers, they should be left as they are, and the quartile deviation should be rounded off to the required accuracy, for this example the nearest mark.

Exercise 28b

1. Find the median and quartile deviation of the following marks: 27, 64, 50, 51, 56, 34, 53, 42, 60, 61, 39, 45, 47.

2. Find the new median and quartile deviation if a fourteenth boy,

who was absent during the examination, obtained 33 marks when he took the paper.

3. Find the upper limit, lower quartile, median and semi-interquartile range of all the numbers divisible by four between zero and 101.

4. Find the median, upper limit, range and quartile deviation of the following table which gives the number of letters in 40 consecutive lines of a column of print.

```
31,  32,  36,  33,  33,  30,  29,  35,  35,  32,
31,  34,  29,  35,  33,  32,  32,  30,  29,  30,
32,  32,  34,  32,  33,  31,  34,  29,  34,  36,
36,  35,  35,  30,  33,  32,  31,  36,  32,  33.
```

28.4

In the above question the numbers are repeated and we could have put down each number with a dot by the side every time that number occurs, and obtain the frequencies of the discrete variable.

(X)	(f)	Cumulative frequencies
29	4	4
30	4	8
31	4	12
32	9	21
33	6	27
34	4	31
35	5	36
36	4	40

$$\Sigma (f)\ 40$$

The dots are evenly spaced and form a frequency diagram. The cumulative frequencies show that the $20\frac{1}{2}$th mark gives a median of 32. The upper quartile is the $10\frac{3}{4}$th mark from the upper end or $30\frac{1}{4}$th from the lower end; this puts it at 34. The lower quartile is 31. The semi-interquartile range is 2 to the nearest whole number.

28.5

We must now consider a continuous variable, such as the heights of 40 boys aged one year. They will be grouped to the degree of accuracy required for a particular purpose. Heights of boys would be grouped to the nearest cm. The class, 60 cm would contain all boys with heights between 59·5 and 60·5 cm. It is very unlikely that anyone would be exactly 60·5 cm, but if they were it would be best to put them in the lower class because of the cumulative frequency. The list of heights would then be formed into a frequency table:

Height (cm) (X):	60	61	62	63	64	65	66	67	68
Frequency (f):	1	0	5	10	9	7	4	2	2

Since it is a continuous variable, the height 65 cm represents a class interval of 1 cm from 64·5 cm, and so a frequency histogram is plotted.

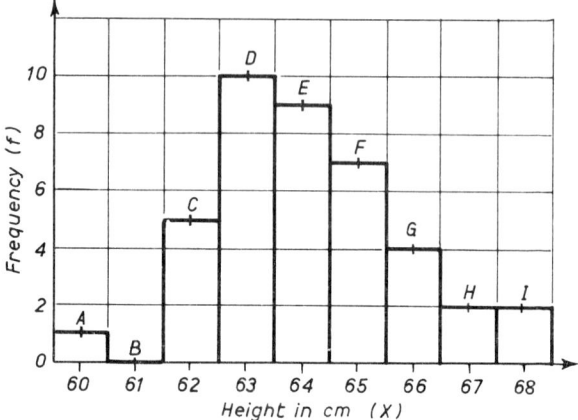

The points A, B, C, D, E, F, G, H, and I can be joined to form a connected point histogram or frequency polygon. It should be used when two distributions are being compared by plotting on the same diagram if there is likely to be confusion due to overlap. The histogram is the best form of representation, as it is the areas of the bars which are propotional to the original frequencies. In our diagram the heights of the bars are also proportional, because the class intervals of X happen to be the same in each case.

The modal class is seen to be 63 cm, and as the original measurements were to the nearest cm, the mode as a single height has no real meaning.

As it is a continuous variable we can use the cumulative frequency curve to estimate the median and quartile deviation. Let us take the first class 60 cm, which includes anyone from just over 59·5 cm up to and including 60·5 cm. ('Just over' can be as small as we like to make it, so the class interval will be 1 cm.) The cumulative frequency 1 must therefore be placed under 'up to and including' 60·5 cm.

	up to								
Height (cm) (X):	60·5	61·5	62·5	63·5	64·5	65·5	66·5	67·5	68·5
Cumulative frequencies:	1	1	6	16	25	32	36	38	40

We now plot a cumulative frequency curve or ogive:

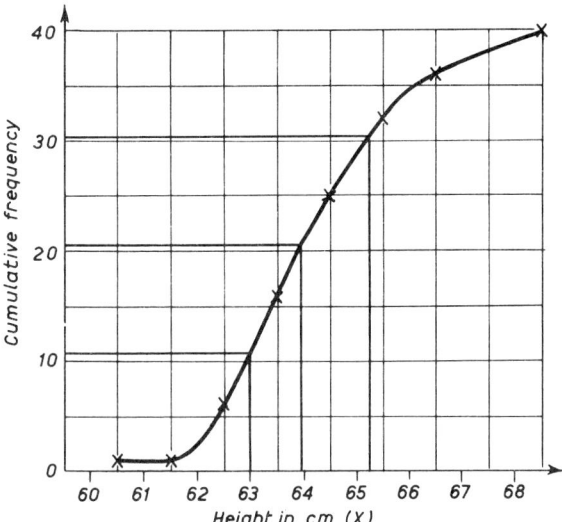

For a total of 40 the median, upper and lower quartiles are the $20\frac{1}{2}$, $30\frac{1}{4}$, and $10\frac{3}{4}$th marks respectively, given by the formulae of section 28.3. These are drawn on to the ogive, dividing it into four equal frequency portions. The values are 63·9, 65·2, and 63·0, giving a median of 63·9 cm and a quartile deviation of 1·1 cm, with an accuracy of $\pm 0·2$ cm (0·2 cm being represented by the side of one square of the graph, and assuming that the original heights were measured to at least this accuracy).

28.6

If we collect our own data we can choose our own class intervals so that they avoid round numbers. For example if we are collecting data on the cost of second-hand cars, using a class interval of £5, it is best to go from, say, £213 to £217 (i.e. $212\frac{1}{2}$ to $217\frac{1}{2}$), then £218–£222, thus keeping £215 and £220 in the middle of a class, because they are quite likely to be the actual prices, as most showrooms round off prices to the nearest £5.

If we are given the data we have to interpret the class interval. The following are some common types of continuously variable class intervals.

(i) 251–300. The full range is from $250\frac{1}{2}$ to $300\frac{1}{2}$ and this gives a range of 50. The cumulative frequency would be for $300\frac{1}{2}$.

Although this will not be noticeable with such large numbers, it would make a difference for small ranges.

(ii) 250–; 300–; 350–; or 250 and less than 300; 300 and less than 350; 350 and less than 400: or 250 and under 300, etc. This can be taken as from 250 to as near to 300 as we like to make it, giving a range of 50. The cumulative frequency would then be plotted just before the 300, but as close as possible to it.

(iii) 4–; 5–; 6– (where 4– means 4 or more but less than 5). We might have got some help from the units but they are not given, although the question states that they are measurements. We must therefore state our own interpretation, the following being possibilities.

(a) 4·000–4·999, class range 1·0, cumulative frequency plotted close to 5.

(b) 3·95–4·95, class range 1·0 cumulative frequency plotted at 4·95. Although the range is from 4 or more, part (b) starts at 3·95, because if the measurement is to the nearest $\frac{1}{10}$ of a unit the difference between 3·95 and 4·00 would not be detected.

(iv) Marks: 0–10; 11–20; 21–30; ...; 91–100. Here the middle groups have a range of 10 (i.e. $10\frac{1}{2}$–$20\frac{1}{2}$), but the first group has a range of $10\frac{1}{2}$ and the last a range of $9\frac{1}{2}$, although they are unlikely to come into any calculations.

(v) Cost in pence: 4–6; 6–8; 8–10. This concerns the cost of making a certain article and so it can be given to three decimal places, as it is probable that the firm work to a cost per 1000 articles. Thus 4–6 should mean a minute amount over 4 to a minute amount under 6, a range of 2 with the cumulative frequency plotted at 6.

If there is any doubt about the class interval you should state your own interpretation.

Exercise 28c. Plot a histogram and an ogive for the following data and estimate the median, quartile deviation and accuracy.

1. Parental Income of University students.

Income	Number of students
Under £250	0
£250 and less than £500	5
£500 ,, ,, ,, £750	20
£750 ,, ,, ,, £1000	43
£1000 ,, ,, ,, £1250	37
£1250 ,, ,, ,, £1500	10
£1500 ,, ,, ,, £1750	5
£1750 ,, ,, ,, £2000	2

2. The following table shows the distribution of shirts in a store according to their price range:

Value	Number of shirts
Under £2·45	0
£2·45 and under £2·65	11
£2·65 ,, ,, £2·85	36
£2·85 ,, ,, £3·05	51
£3·05 ,, ,, £3·25	63
£3·25 ,, ,, £3·45	47
£3·45 ,, ,, £3·65	18
£3·65 ,, ,, £3·85	12
£3·85 ,, ,, £4·05	5

3. Measurement: 6–, 7–, 8–, 9–, 10–, 11–, 12–, 13–
   Frequency: 4 7 17 20 28 16 6 1
   (The interval 6– means 6 or more but less than 7.)

4. The heights of a group of 33 shrubs were distributed as follows:

Height in cm:	57	58	59	60	61	62	63
Frequency:	1	5	9	10	5	2	1

5. The following are the absences per term for a school taken in chronological order: 195, 164, 91, 162, 183, 107, 152, 143, 120, 120, 115, 90, 133, 134, 106, 142, 132, 97, 227, 234, 120, 173, 153, 111, 183, 114, 90, 176, 133, 121, 100, 97, 90. (Take class intervals 90–, 120–, 150–, 180–, and 210– to form frequency table.)

6. The cost of production of a packet of 20 cigarettes by different makers is shown in the following data;

Cost in pence:	6–8	8–10	10–12	12–14	14–16	16–18
Number of makers:	1	15	29	14	2	1

28.7

Very often data is collected with unequal class intervals, although we should avoid this if carrying out the collection of data ourselves. A histogram can be plotted if we remember that it is the area of the bars which are proportional to the frequencies. Let us take a table showing the frequencies of primary schools into a county according to the number of pupils.

	up to					above
No. of pupils (X):	25	26–50	51–100	101–150	151–250	251
No. of schools (f):	1	6	25	59	60	7

It is reasonable to assume that every primary school has at least 16 pupils and that they are unlikely to have more than 300. If we take the third and fourth classes as standards we have the first group as a fifth, the second as half and the fifth as twice the standard class interval (Notice that the interval 51–100 is 50 pupils, as the numbers are inclusive.)

If we take the standard class as a single base for the histogram the first group has only a fifth base and so its frequency must be mutiplied by five to keep the areas equivalent. Similarly the second frequency must be doubled and the fifth frequency halved.

No. of
pupils (X): 16–25   26–50   51–100   101–150   151–250   251–300

Frequency
for
standard
class: 5   12   25   59   30   7

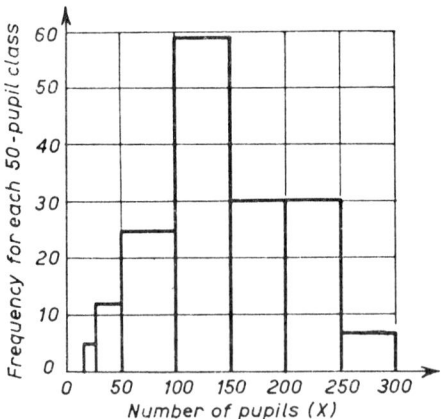

The frequency is stated for the standard class when the histogram is drawn. The areas are now proportional to the frequencies. The cumulative frequencies are obtained from the original table of data.

Exercise 28d. Plot a histogram and an ogive for the following data and estimate the median, quartile deviation and accuracy.

1. Intelligence quotients of children in a secondary modern school:

I.Q.:	75–84	85–89	90–94	95–99	100–104	105–114	115–124
Frequency:	60	42	45	39	27	33	4

2. The following is an analysis of cinemas by size:

Seating capacity	No. of cinemas
Under 250	2
251–500	67
501–750	124
751–1000	129
1001–1500	82
1501–2000	23
2001–3000	16

3. Analysis of grocery shops by floor area in tens of square metres.

Area (10's m²):	1–3	4–6	7–12	13–21	22–30
No. of shops:	24	37	28	18	6

4. Analysis of results in Mathematics examination.

Percentage:	0–29	30–39	40–49	50–54	55–59	60–69	70–79	80–99
No. of pupils:	43	45	106	80	67	115	95	76

28.8

The median and quartiles can be calculated quickly if we add a cumulative frequency column to the given data. We shall do this for the example of section 28.5:

Height (cm) (X):	60	61	62	63	64	65	66	67	68
Frequency (f):	1	0	5	10	9	7	4	2	2
Class end (cm):	60·5	61·5	62·5	63·5	64·5	65·5	66·5	67·5	68·5
Cumulative frequencies (cf):	1	1	6	16	25	32	36	38	40

For a total of 40 boys the median is the $20\frac{1}{2}$th value. We see that 16 boys are 63·5 cm or under. The class range of 1·00 cm of the next class contains 9 boys, and we assume that they are evenly spaced. We have to go $(20\frac{1}{2} - 16)$th of the way into this class, i.e. $4\frac{1}{2}$. The extra height is $\frac{4\frac{1}{2}}{9}$. 1·00 cm $= 0·5$ cm. The median is 63·5 cm $+ 0·5$ cm $= 64·0$ cm. This answer will have an accuracy of about one-fifth of a class, i.e. 0·2 cm.

The lower quartile is the $10\frac{3}{4}$th value. There are 6 boys of 62·5 cm or under, and 10 boys in the next class. We have to go $(10\frac{3}{4} - 6)$th of the way into the class. $\frac{4\frac{3}{4}}{10}$. 1·00 cm $= 0·48$ cm. The lower quartile is 62·5 cm $+ 0·5$ cm $= 63·0$ cm.

The upper quartile is the $30\frac{1}{4}$th value. There are 25 boys of 64·5 cm or under, and 7 boys in the next class. We have to go $5\frac{1}{4}$th of the way into the class. $\dfrac{21}{4.7} \cdot 1\cdot00$ cm $= 0\cdot75$ cm. The upper quartile is

$64\cdot5$ cm $+ 0\cdot8$ cm $= 65\cdot3$ cm.

The quartile deviation is $\frac{1}{2}(65\cdot3 - 63\cdot0) = 1\cdot2$ cm (to $0\cdot1$ cm).

This gives a median of $64\cdot0$ cm and quartile deviation of $1\cdot2$ cm (each $\pm 0\cdot2$ cm), which compare with the graphical method of section 28.5.

Exercise 28e. Calculate the medians and quartile deviations of Questions 1–6 of Exercise 28c and Questions 1–4 of Exercise 28d (number the last four, 7–10).

29. Mean, Standard and Mean Deviations

29.1

Although the mode and median can be found easily they are not as useful averages as the arithmetical mean. It is defined as the sum of the numbers divided by the number of numbers. Thus if we have N numbers X_1, X_2, X_3, X_4, ... X_N the mean $\bar{X} = \dfrac{\Sigma X_N}{N}$

To find the mean of the integers 1 to 10

$$\Sigma X_N = 1+2+3+4+5+6+7+8+9+10 = 55$$

$$N = 10$$

The mean $\bar{X} = \dfrac{55}{10} = 5\cdot5$.

29.2

The range and the quartile deviation have been used as measures of dispersion of the marks but the most comprehensive is the standard deviation. It is the root mean square deviation so that

$$s = \sqrt{\dfrac{\Sigma (X - \bar{X})^2}{N}}$$

If we are to find the standard deviation of a set of numbers we calculate the mean $\bar{X}$ by the method of the previous section. The difference between each number and the mean is found $(X - \bar{X})$, this is called the deviation from the mean. The deviation is squared

$(X - \bar{X})^2$ and these squares are summed $\sum (X - \bar{X})^2$. The sum is then divided by the total of numbers N and the square root gives the standard deviation.

To find the standard deviation of the integers 1 to 10, having found the mean 5·5.

X	$(X - \bar{X})$	$(X - \bar{X})^2$
1	$-4\cdot5$	20·25
2	$-3\cdot5$	12·25
3	$-2\cdot5$	6·25
4	$-1\cdot5$	2·25
5	$-0\cdot5$	0·25
6	0·5	0·25
7	1·5	2·25
8	2·5	6·25
9	3·5	12·25
10	4·5	20·25
	0	82·50

As a check the sum of the deviations is zero.

$$\sum (X - \bar{X})^2 = 82\cdot5 \quad \text{and} \quad N = 10$$
$$\therefore \frac{\sum (X - \bar{X})^2}{N} = 8\cdot25$$

this is s^2 and is called the variance.

$$s = \sqrt{8\cdot25} = 2\cdot87$$

29.3

The coefficient of variation (v) gives us a method of comparing the scatter of two or more different sets of values. This is found by expressing the standard deviation as a percentage ratio of the mean. Since the mean and standard deviation are measured in the same units the variance is a ratio independent of units. For the integers 1 to 10:

$$v = \frac{s}{\bar{X}}.100 = \frac{2\cdot87}{5\cdot5}.100 = 52\cdot2\%$$

The greater this figure the more the numbers are dispersed about the mean.

29.4

A boy obtains 72% for English and 76% for Mathematics. How much information is needed to decide which is the better mark? We need to know the mean of all the English marks and their standard deviation, they are 60 and 12. For the Mathematics mark they are 60 and

16 respectively. To compare we have to reduce each to a standard score (S.S.) which equates each mean to zero and each standard deviation to 1.

$$\text{Standard score (S.S.)} = \frac{\text{deviation}}{\text{standard deviation}} = \frac{(X - \bar{X})}{s}$$

English:

$$\text{S.S.} = \frac{72 - 60}{12} = +1$$

Mathematics:

$$\text{S.S.} = \frac{76 - 60}{16} = +1$$

The boy's standard is the same in each subject. There is no need to reduce marks to a percentage to find the standard score. It compensates for the wider scatter of Mathematics marks about the mean and any other differences. Full marks or no marks are often given for Mathematics but most English masters reserve full marks for the perfect essay, which they do not expect to receive from pupils.

29.5

The mean deviation can be found from the mode, median or mean, but is a minimum when found from the median. It is no longer used in modern practical work as the standard deviation is a much better measure of dispersion. Since its size is usually compared with the standard deviation we shall calculate it from the mean, but from the alternatives if they are easier.

$$\text{Mean deviation} = \frac{\sum |(X - \bar{X})|}{N}$$

$(X - \bar{X})$ is the deviation and $|(X - \bar{X})|$ is called modulus $(X - \bar{X})$, which is the deviation ignoring any negative signs. Thus

$$|p| = |-p|$$

Taking the integers 1 to 10 with a mean of 5·5:

X:	1	2	3	4	5	6	7	8	9	10
$(X - \bar{X})$:	−4·5	−3·5	−2·5	−1·5	−0·5	0·5	1·5	2·5	3·5	4·5
$|(X - \bar{X})|$:	4·5	3·5	2·5	1·5	0·5	0·5	1·5	2·5	3·5	4·5

$$\sum |(X - \bar{X})| = 25, \quad N = 10$$

$$\therefore \text{Mean deviation} = 2\cdot5$$

(In this problem the mean and the median are the same, but they will not be in most frequency distributions.)

Exercise 29a

1. Calculate the mean and standard deviation for the integers 11 to 19. What is the mean deviation?

2. In a plantation of Christmas trees the mean height is 2·06 m and the standard deviation 0·45 m. In a second the mean height is 1·18 m and the standard deviation 0·35 m. Compare the percentage variation.

3. Calculate the mean and standard deviation for the integers between 1 and 20 which are not divisible by 3 or 4. Find the mean deviations from the mean and the median.

4. Find the standard scores for the following examination marks:
   English: 189 out of 300, standard deviation 36, mean 150.
   Mathematics: 136 out of 200, standard deviation 30, mean 100.
   French: 62% standard deviation 16, mean 43.
   Latin: 69% standard deviation 9, mean 60.

5. Find the mean, standard, and mean deviations of the numbers 6, 8, 10, 12, 14.

6. Find the mean, standard, and mean deviations of the numbers 12, 16, 20, 24, 28.

29.6

With large or decimal numbers we can use the assumed mean, which is a rough estimate of the mean. It can be used to find the correct mean and in place of the correct mean to find the standard and mean deviations.

Consider the numbers 50·6, 50·8, 50·9, 51·1 and 51·4. A rough estimate of the mean is 51 and we shall take this as the assumed mean (A).

X	A	$(X-A)$	$(X-A)^2$	$|(X-A)|$
50·6	51·0	−0·4	0·16	0·4
50·8	51·0	−0·2	0·04	0·2
50·9	51·0	−0·1	0·01	0·1
		——		
		−0·7		
		——		
51·1	51·0	+0·1	0·01	0·1
51·4	51·0	+0·4	0·16	0·4
		——	——	——
		+0·5	0·38	1·2
		−0·7		
		——		
		−0·2		

If A had been the correct mean, $\sum (X - A)$ would have been zero. The difference -0.2 must be five times the difference between A and $\bar{X}$:

$$\frac{\sum (X-A)}{N} = \frac{-0.2}{5} = -0.04 \quad \text{(this is a discrepancy)}$$

The mean

$$\bar{X} = A + \frac{\sum (X-A)}{N} = 51.00 - 0.04 = 50.96$$

$$s = \sqrt{\frac{\sum (X-A)^2}{N} - \left(\frac{\sum (X-A)}{N}\right)^2} = \sqrt{\frac{0.38}{5} - (-0.04)^2}$$

$$= \sqrt{0.076 - 0.0016}$$

$$= \sqrt{0.0744} = 0.273$$

To find the mean deviation from the mean each deviation in the $(X - A)$ column is 0.04 too small compared with $(X - \bar{X})$. When we take the modulus, those with a positive sign in the $(X - A)$ column must have 0.04 added, i.e. 2×0.04. Those with a negative sign in the $(X - A)$ column have their signs changed and so each must have 0.04 subtracted, i.e. -3×0.04. The net correction is -0.04.

$$\therefore \text{ mean deviation from the mean} = \frac{1.2 - 0.04}{5} = \frac{1.16}{5} = 0.232$$

The working applies only if there are no X numbers between A and $\bar{X}$. If there are, a $|(X - \bar{X})|$ column must be calculated and summed as in section 29.5.

To find the mean deviation from the median 50.9:

X:	50·6	50·8	50·9	51·1	51·4
$|(X-\text{median})|$:	0·3	0·1	0	0·2	0·5

$$\text{Mean deviation from the median} = \frac{1.1}{5} = 0.22$$

Exercise 29b

1. Find the mean and standard deviation of the following numbers: 21·6, 21·8, 22·0, 22·1, 22·2, 22·3, 22·5, 22·6, 22·7, 23·0.

2. Find the mean, standard deviation and mean deviation from the median of the following numbers: 126, 129, 134, 136, 137, 138, 139, 140, 140.

3. Find the mean, mean deviations from the median and the mean and the standard deviation of the following numbers: 926, 937, 950, 966, 970, 981, 999, 1003.

29.7

With a frequency distribution it is only necessary to remember that it is made up of repeated numbers and so the deviation, deviation squared, and modulus deviation columns have to be multiplied by the frequency. Let us take the problem of section 28.1. The assumed mean is 40.

Number of sticks (1) X	Number of boxes (2) f	(3) $(X-A)$	(4) $f.(X-A)$	(5) $f.(X-A)^2$	(6) $|f.(X-A)|$
36	3	-4	-12	48	12
37	14	-3	-42	126	42
38	10	-2	-20	40	20
39	36	-1	-36	36	36
40	73	0	-110	0	0
41	27	$+1$	27	27	27
42	20	$+2$	40	80	40
43	17	$+3$	51	153	51
	200		118 −110	510	228
			8		

For each row:

Column 3 is the difference between the number and the assumed mean.

Column 4 is column 3 multiplied by column 2.

Column 5 is column 4 multiplied by column 3, and checked by squaring column 3 and multiplying by column 2.

Column 6 is the modulus of column 4, and its total is checked by adding 118 and 110, the sub-totals of column 4.

$$\Sigma f = N = 200$$

$$\Sigma f.\frac{(X-A)}{N} = \frac{8}{200} = 0.04$$

Mean $= 40 + 0.04 = 40.04$

$$s = \sqrt{\frac{510}{200} - (0.04)^2} = \sqrt{2.55 - 0.0016}$$

$$= \sqrt{2.548} = 1.60$$

For the mean deviation from the mean we see that each value of the column $(X-A)$ is 0·04 too big, compared with $(X-\bar{X})$. When we take the modulus those with a positive sign must have 0·04 subtracted, i.e. $-(27+20+17).0.04$. Those with a negative sign in the column $(X-A)$, and also the zero of row 5 (because it is -0.04 from true mean), all have to have 0·04 added, i.e. $(3+14+10+36+73).0.04$. The net correction is $(136-64).0.04 = 72.0.04 = 2.88$.

$$\text{Mean deviation from the mean} = \frac{228+2.88}{200} = 1.15$$

Since the array is discrete the mode must be 40 sticks, giving a mean deviation from the mode of $\frac{228}{200} = 1.14$. The median is also 40, giving the same mean deviation. For continuous variables it should be unnecessary to work from mode or median.

Exercise 29c

1. Find the mean and standard deviation of the following table, showing the frequency of letters in 40 lines of print.

No. of letters (X):	29	30	31	32	33	34	35	36
No. of lines (f):	4	4	4	9	6	4	5	4

2. Find the mean height and standard deviation for the following table of baby boys' heights.

Height (cm):	60	61	62	63	64	65	66	67	68
Frequency:	1	0	5	10	9	7	4	2	2

3. Find the mean and standard deviation of the number of beans in each of 100 packets.

No. of beans:	24	25	26	27	28	29	30	31	32	33
No. of packets:	6	8	9	12	18	21	12	8	0	6

Calculate the mean deviations from the mean and from the median.

4. Find the mean and standard deviation of the number of times that 303 children stayed to school lunches in a period of 2 weeks (10 school days):

No. of meals:	0	1	2	3	4	5	6	7	8	9	10
No. of children:	68	8	5	0	23	50	8	10	32	42	57

5. Find the mean, standard deviation and mean deviation (from the mean) of the following data concerning the selling price of a tin of coffee by 62 different shops.

Cost in pence:	25	26	27	28	29	30
Number of shops:	1	15	29	14	2	1

6. The lengths of a 100 rods were measured with the following results:

Length (to nearest $\frac{1}{10}$ cm): 5·6 5·7 5·8 5·9 6·0 6·1 6·2
Frequency: 3   9   28   29   20   10   1

Calculate the mean, standard deviation, and mean deviation from the mean.

7. The numbers of faults on the surface of each of 1000 tiles were distributed as shown:

No. of faults:	0	1	2	3	4	5
Frequency:	760	138	67	25	8	2

Calculate the mean and standard deviation.

29.8

When a continuous variable is grouped into equal or unequal class intervals the working can be simplified by considering each class to be represented by its mean (X').

The following are the heights of Christmas trees measured to the nearest cm.

Height (X)	f	X'	x	fx	fx^2	$|fx|$
43–45·5 cm	1	44·25	−12	−12	144	12
46–48·5 cm	2	47·25	−9	−18	162	18
49–51·5 cm	6	50·25	−6	−36	216	36
52–54·5 cm	21	53·25	−3	−63	189	63
55–57·5 cm	44	56·25	0	−129	0	0
58–60·5 cm	23	59·25	+3	69	207	69
61–63·5 cm	3	62·25	+6	18	108	18
	100			87	1026	216
				−129		
				−42		

$$\Sigma f = N = 100$$

$$\frac{\Sigma fx}{N} = \frac{-42}{100} = -0.42$$

$$s = \sqrt{\frac{\Sigma f.x^2}{N} - \left(\frac{\Sigma fx}{N}\right)^2} = \sqrt{\frac{1026}{100} - (-0.42)^2}$$

$$= \sqrt{10.26 - 0.18}$$

$$= \sqrt{10.08}$$

$$= 3.17$$

The discrepancy $= -0.42$ cm.
The mean $\overline{X}$ is 56·25–0·42 cm $= 55·83$ cm.
The mean $= 55·8$ cm.
The standard deviation $s = 3·17$ cm.

For the mean deviation from the mean we see that the deviation x of the zero class (in row 5) is 0·42 cm too small, compared with the deviation from the true mean.

When we take the modulus, the positive deviations and zero must have 0·42 added, i.e. $(44 + 23 + 3).0·42$. The negative deviations must have 0·42 subtracted, i.e. $-(1 + 2 + 6 + 21).0·42$. The net correction is $(70 - 30).0·42 = 40 \times 0·42 = 16·8$.

$$\text{Mean deviation} = \frac{216 + 16·8}{100} \text{ cm}$$

$$= 2·328 \text{ cm}$$

$$\text{Mean deviation from the mean} = 2·33 \text{ cm}$$

It is seen that if there are no X groups between A and $\overline{X}$ the following rule applies for the mean deviation correction. The frequencies of groups above the zero (guessed mean) group are taken as positive and those below are taken as negative. If the discrepancy is positive the zero group frequency is taken as positive, but if negative the zero group frequency is taken as negative. The net sum is then multiplied by the discrepancy to give the correction to the sum of the modulus deviation.

For the above example:

$$\text{Correction} = (1 + 2 + 6 + 21 - 23 - 3 - 44).(-0·42)$$

$$= (-40).(-0·42) = +16·8$$

Exercise 29d. Calculate the mean, standard deviation, and mean deviation from the mean for Questions 1–6 of Exercise 28c.

30. Moving Averages; Index Numbers

30.1

If we plot a time series, such as the amount of quarterly electricity bills, it will be noticed that there is a variation over the four quarters. This is because the summer quarter is lighter and warmer, and so less electricity is used. If we wish to find the general trend from year to year we must try and smooth out these variations. This is done by plotting the yearly moving averages. If over three years the values of the twelve bills are x_1, x_2, x_3, x_4, x_5, x_6, x_7, x_8, x_9, x_{10}, x_{11}, and x_{12}, the first four-quarterly moving average

$$m_1 = \frac{x_1 + x_2 + x_3 + x_4}{4}$$

The second:

$$m_2 = \frac{x_2 + x_3 + x_4 + x_5}{4}$$

The final:

$$m_9 = \frac{x_9 + x_{10} + x_{11} + x_{12}}{4}$$

The points are plotted at the mid-intervals, i.e. m_1 above the $2\frac{1}{2}$th quarter, and should smooth out the variations to show the general trend as an increase, steady, or a decrease.

The range of the moving average depends on the data. If it is concerned with the school week it will be a five-day moving average. Most shops are open six days a week and would have a six-day moving average. If monthly figures are given there will be variations due to the different lengths of month as well as seasonal changes (e.g. in industry). Here we should use a twelve-monthly moving average. Finally much natural data changes due to cyclical variations over five or ten years.

A convenient way of finding m_2 is to notice that one number, $x_{(n+1)}$, is added and one, x_1, dropped. The change must be shared over n numbers.

If

$$m_1 = \frac{x_1 + x_2 + \ldots + x_n}{n}$$

then

$$m_2 = m_1 + \frac{x_{(n+1)} - x_1}{n}$$

and $$m_3 = m_2 + \frac{x_{(n+2)} - x_2}{n}, \text{ etc.}$$

As a check the final n numbers should be added together and divided by n, and should equal the final moving average.

The following are the quarterly electricity bills for three years for a school (in pounds to the nearest pound).

Quarter:	1	2	3	4
1968:	87	58	48	102
1969:	149	60	48	86
1970:	154	114	58	160

Plot them and also the four-quarterly averages.

Differences (Note original table is set out in four quarters):

2nd − 1st row:	62	2	0	− 16
3rd − 2nd row:	5	54	10	74

$m_1 = \frac{1}{4}(87+58+48+102) = \frac{1}{4}(295) \quad = \quad 73\cdot75 \quad$ at $2\frac{1}{2}$ qr.

$m_2 = \quad 73\cdot75 + \dfrac{62}{4} \quad = \quad 73\cdot75 + 15\cdot5 \quad = \quad 89\cdot25 \quad$ at $3\frac{1}{2}$

$m_3 = \quad 89\cdot25 + \dfrac{2}{4} \quad = \quad 89\cdot25 + \ 0\cdot5 \quad = \quad 89\cdot75 \quad$ at $4\frac{1}{2}$

$m_4 = \quad 89\cdot75 + \dfrac{0}{4} \quad = \quad 89\cdot75 + \ 0 \quad = \quad 89\cdot75 \quad$ at $5\frac{1}{2}$

$m_5 = \quad 89\cdot75 + \dfrac{(-16)}{4} \quad = \quad 89\cdot75 - \ 4\cdot0. \quad = \quad 85\cdot75 \quad$ at $6\frac{1}{2}$

$m_6 = \quad 85\cdot75 + \dfrac{5}{4} \quad = \quad 85\cdot75 + \ 1\cdot25 \quad = \quad 87\cdot0 \quad$ at $7\frac{1}{2}$

$m_7 = \quad 87\cdot0 \ + \dfrac{54}{4} \quad = \quad 87\cdot0 \ + 13\cdot5 \quad = \quad 100\cdot5 \quad$ at $8\frac{1}{2}$

$m_8 = 100\cdot5 \ + \dfrac{10}{4} \quad = \quad 100\cdot5 \ + \ 2\cdot5 \quad = \quad 103\cdot0 \quad$ at $9\frac{1}{2}$

$m_9 = 103\cdot0 \ + \dfrac{74}{4} \quad = \quad 103\cdot0 \ + 18\cdot5 \quad = \quad 121\cdot5 \quad$ at $10\frac{1}{2}$

Check: $m_9 = \frac{1}{4}(154+114+58+160) = \frac{1}{4}(486) = 121\frac{1}{2}$

The moving averages are plotted as dots.

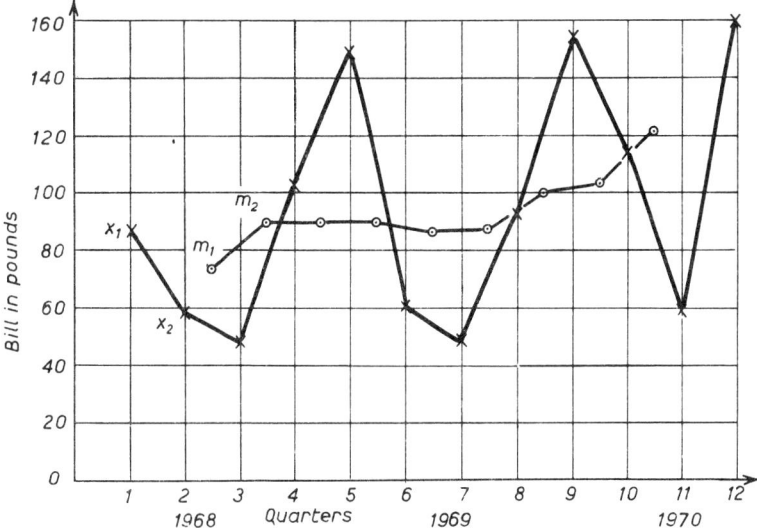

The general trend shows a steady period and then a rise. The school changed over from coal to electrical heating in February 1970.

Exercise 30a. Plot the following data and their moving averages for the period indicated.

1. Electricity bills for 1966–8 per quarter to the nearest pound. (Notice that although the x values overlap the example, the m values do not.)

Quarter:	1	2	3	4
1966:	56	43	45	81
1967:	106	50	46	51
1968:	87	58	48	102

(Four-quarterly moving averages)

2. Quarterly gas bills for 1967–70 in pounds to the nearest pound.

Quarter:	1	2	3	4
1967:	39	47	25	56
1968:	68	59	66	72
1969:	88	60	60	67
1970:	76	70	56	75

(Four-quarterly moving averages)

3. Monthly motor-boat production:

Month:	Jan.	Feb.	Mar.	Apr.	May	June
1969:	64	60	51	46	36	30
1970:	60	55	49	46	36	33

Month:	July	Aug.	Sept.	Oct.	Nov.	Dec.
1969:	29	32	38	39	48	54
1970:	31	33	38	38	47	54

(Twelve-monthly moving averages, work in fractions not decimals)

4. Sales of new motor cars by one showroom:

Year:	1958	1959	1960	1961	1962	1963
Sales:	1517	1498	1451	1536	1690	1738

Year:	1964	1965	1966	1967	1968	1969
Sales:	1892	2017	2116	1934	1900	2107

(Three-yearly moving averages)

5. Daily absences from school during three weeks.

Week:	1	2	3
Monday:	23	38	52
Tuesday:	28	52	54
Wednesday:	21	43	61
Thursday:	33	58	51
Friday:	40	63	51

(Five-day span excluding Saturday and Sunday)

6. Number of days on which the sun shone for more than 2 hours in each month.

Month:	Jan.	Feb.	Mar.	Apr.	May	June
1968:	3	6	8	11	16	21
1969:	3	8	13	17	21	24
1970:	15	17	20	18	20	21

Month:	July	Aug.	Sept.	Oct.	Nov.	Dec.
1968:	24	26	25	18	7	0
1969:	26	22	14	10	11	13
1970:	23	24	16	12	12	10

30.2

One of the most important index numbers is the index of retail prices, which is published monthly. It started in 1904 as the cost-of-living index and its basis was revised at various times until 1947. It was found after the war that the method of calculation was out of date, due to the changing way of life, because of the use of the internal combustion engine, electricity and many other factors. The interim index of retail prices was introduced in 1947 and the index of retail prices was introduced in January 1956. The latest base year is 1964.

Each monthly index number is an attempt to estimate the cost of running a house for that month, compared with the cost in January 1964.

Spending is divided into ten groups for an average household. In January 1956 the amount spent in each group was made equivalent to 100, which is the price relative and corresponds to the (X) column of section 29.7.

The first group is food and the second alcoholic drink. A greater proportion of the household budget is spent on food than on drink, and so an increase of 5 % in the cost of food would make a greater difference than 5 % in the cost of drink. Thus food must play a greater part in the calculations. This is done by giving each group a weight which corresponds to the frequency column of section 29.7.

JANUARY 1956 INDEX OF RETAIL PRICES

Group	Price relative (X)	Weight (w)
Food	100	350
Alcoholic drink	100	71
Tobacco	100	80
Housing	100	87
Fuel and light	100	55
Durable goods	100	66
Clothing	100	106
Transport	100	68
Other goods	100	59
Services	100	58

The index of retail prices is the weighted arithmetical mean and is calculated by working out a third column wX and summing it. The index $= \dfrac{\sum wX}{\sum w}$, and the weights add up to a 1000. The index for January 1956 is 100.

The groups are each split up into sub-groups according to a schedule of the Ministry of Labour. Each item within the group is weighted in order to obtain the price relative. For any particular month the price relatives may be different from 100, but of course the weights remain the same. The calculations are simplified if we take 100 as the assumed mean, which brings out the relative changes of the groups.

CALCULATION OF INDEX OF RETAIL PRICES FOR JANUARY
1961

(January 1956 was 100)

Group	Price relative (X)	Deviation (X−A)	Weight w	w(X−A)
Food . . .	107·7	7·7	350	2695·0
Drink . . .	98·4	−1·6	71	−113·6
Tobacco . . .	113·1	13·1	80	1048·0
Housing . . .	134·0	34·0	87	2958·0
Fuel and light . .	125·8	25·8	55	1419·0
Durable goods . .	99·4	−0·6	66	−39·6
Clothing . . .	104·8	4·8	106	508·8
Transport . .	120·4	20·4	68	1387·2
Other goods . .	118·5	18·5	59	1091·5
Services . . .	122·8	22·8	58	1322·4
			1000	12,429·9
				−153·2
				12,276·7

$$\text{Index} = 100 + \frac{\sum w(X-A)}{\sum w}$$

$$= 100 + \frac{12,276\cdot7}{1000}$$

$$= 112\cdot3$$

The index is worked out to the first decimal place, but is usually quoted to the nearest whole number. If prices are given for two years, the price relatives for the base year are taken as a 100 each, and the other year's price relatives are found by ratio. For example if eggs are 2p in the base year, and then 3p, the second year's price relative would be 150.

Exercise 30b

1. Calculate the cost-of-living index from the following table:

	Weight	Price relative
Food . . .	40	109
Rent . . .	7	103
Clothes . . .	10	96
Fuel . . .	6	101
Tobacco and drink .	17	100
Other goods, etc. .	20	99
	100	

2. Calculate the industrial production index of the engineering and allied industries for January 1961 (1954 as base).

	Weight	Index
Engineering . .	164	124
Shipbuilding . .	22	91
Vehicles . . .	78	131
Metal goods . .	42	127
	306	

3. Find the price relative for the sub-group Food for January 1961 (1956 base 100).

	Weight	Price relative
Bread, cakes, etc. .	52	123
Meat and bacon .	89	110
Fish . . .	9	122
Butter and fats . .	19	84
Milk, cheese, and eggs	53	111
Tea, coffee, etc. .	22	98
Sugar and preserves .	39	104
Vegetables . .	33	98
Fruit . . .	19	100
Other food . .	15	105
	350	

4. Calculate an index number for the second year, taking the first as base. (Prices in pence per kg.)

	Beef	Lamb	Pork	Sausages, pies, poultry, etc.
1st year	63	70	72	60
2nd year	84	75	74	55
Weight	20	15	25	40

5. Calculate the index of retail prices for January 1960 (January 1956 was 100).

Group	Price relative	Weight
Food	107·8	350
Alcoholic drink	98·1	71
Tobacco	108·1	80
Housing	129·3	87
Fuel and light	119·0	55
Household goods	97·6	66
Clothing	103·0	106
Transport and vehicles	116·0	68
Other goods	113·9	59
Services	117·9	58
		1000

6. Find the latest index of retail prices using the table published by H.M.S.O. in *The Monthly Digest of Statistics*.

31. Probability; Significance

31.1

If an event can happen in n equally likely ways, and a special feature occurs in r of those ways, then the probability or chance of the special event $p = \dfrac{r}{n}$

It is difficult to be sure that the n events are equally likely. In horse racing it is the unequal possibilities which need to be judged when the bookmaker lays the odds. The serious gambler thinks he is the better judge and so bets on a horse which is often called a certainty. In mathematics the only certainty is $p = 1$. This would mean only one horse in the race, for if there are two the second must stand some chance, even if it is very remote. The chance of the second horse winning will never be an impossibility, i.e. $p = 0$. In many cases we assume that the n events are equally likely, as with throwing a normal die or tossing a normal coin.

Let us suppose that a car park contains 50 cars and it is equally likely that any one may leave. If there are 10 Fords the probability that the first to leave is a Ford (p) is $\dfrac{10}{50}$, i.e. $\dfrac{1}{5}$.

31.2

If two events with probabilities p_1 and p_2 are mutually exclusive then the probability that *either* one *or* the other occurs is the sum of their separate probabilities $p_1 + p_2$.

The park has an exit which will allow only one car to pass. When that car is passing none of the others can and so the event is mutually exclusive. If there are 7 Vauxhalls then; using the tree diagram:

For the Fords:

$$p_1 = \frac{r_1}{n_1} = \frac{10}{50}$$

For the Vauxhalls:

$$p_2 = \frac{r_2}{n_2} = \frac{7}{50}$$

Therefore the *total* probability that *either* a Ford *or* a Vauxhall leaves first (shown by branches A and B) is $p_1 + p_2$, i.e. $\dfrac{17}{50}$.

31.3

If two events with probabilities p_1 and p_2 are independent, then the probability that *both* the one *and* the other occur is the product of their separate probabilities, $p_1 \times p_2$.

Suppose that a second car park has six cars, one of which is a Ford. Then a car leaving from one park is independent of any changes in the other.

In the first park for a Ford:

$$p_1 = \frac{r_1}{n_1} = \frac{10}{50} = \frac{1}{5}$$

In the second park for a Ford:

$$p_2 = \frac{r_2}{n_2} = \frac{1}{6}$$

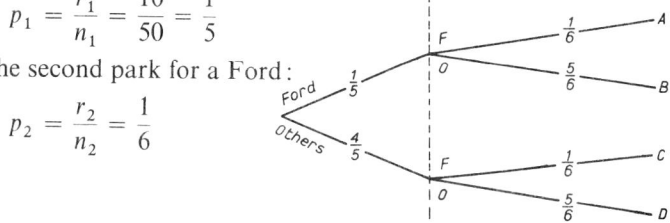

Therefore the *combined* probability that a Ford will be the first to leave from *both* the first *and* second parks (branch A) is $p_1 . p_2$, i.e. $\dfrac{1}{30}$

31.4

If an event has a probability of p, the probability that it will not happen is $(1 - p)$ or q.

To find the probability that the first car will be a Ford from the first car park, but another make from the second (branch B above)

$$p_1 = \frac{1}{5}; \qquad q_2 = (1 - p_2) = \frac{5}{6}$$

The probability is $\qquad\qquad \dfrac{1}{5} . \dfrac{5}{6} = \dfrac{1}{6}$

31.5

When a second event follows a first the probability of the second event will depend on what has happened during the first.

We shall find the probability that the first car from the first park is a Ford and the second a Vauxhall; drawing only the branches needed.

For the first car to be a Ford: $\quad p_1 = \dfrac{r_1}{n_1} = \dfrac{10}{50} = \dfrac{1}{5}$

At this stage one car has left the park and so $n_2 = 49$.

For the second car to be a Vauxhall: $\quad p_2 = \dfrac{r_2}{n_2} = \dfrac{7}{49} = \dfrac{1}{7}$

Combined probability (branch B) $= \dfrac{1}{35}$

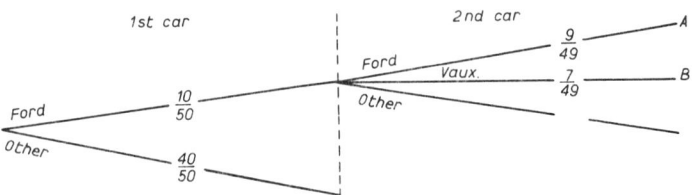

31.6

The rules which have been applied to two events can be extended to three or more, and can be used in conjunction with each other.

Let us find the probability that the first car from the first park is not a Ford, the second is a Vauxhall and the third is a Ford. The problem will have to be considered in two parts as the first car may or may not be a Vauxhall.

(i) First car a Vauxhall: $p_1 = 7/50$

Second car a Vauxhall: $p_2 = 6/49$

Third car a Ford: $p_3 = 10/48$

Combined probability (branch A)

$$\frac{\overset{1}{\cancel{7}}}{\underset{5}{\cancel{50}}} \cdot \frac{\overset{1}{\cancel{6}}}{\underset{7}{\cancel{49}}} \cdot \frac{\overset{1}{\cancel{10}}}{\underset{8}{\cancel{48}}} = \frac{1}{5.7.8}$$

(ii) If the first car is either a Vauxhall or a Ford, we found $p_1 = 17/50$

∴ First car neither Ford nor Vauxhall: $q_1 = 33/50$

Second car a Vauxhall: $p_2 = 7/49$

Third car a Ford: $p_3 = 10/48$

Combined probability (branch D)

$$\frac{\overset{11}{\cancel{33}}}{\underset{5}{\cancel{50}}} \cdot \frac{\overset{1}{\cancel{7}}}{\underset{7}{\cancel{49}}} \cdot \frac{\overset{1}{\cancel{10}}}{\underset{16}{\cancel{48}}} = \frac{11}{5.7.16}$$

The total probability of (i) and (ii) is

$$\frac{1}{5.7.8} + \frac{11}{5.7.16} = \frac{13}{35.16} = \frac{13}{560}$$

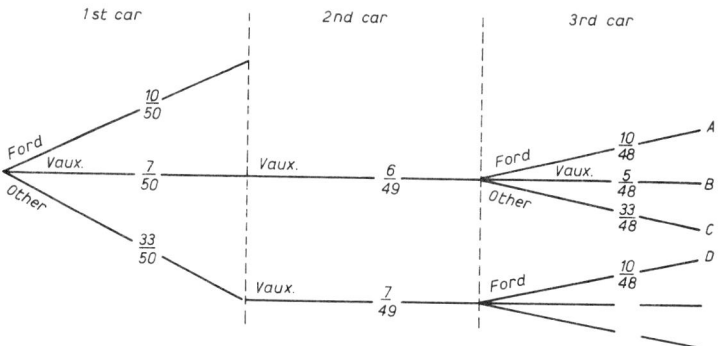

Exercise 31a. (Assume events are equally likely.)

1. A coin is tossed at the same time as a playing card is drawn. What is the chance of (i) a head and a diamond, (ii) a head and an ace, (iii) a head and the ace of diamonds?

2. A die is thrown and a playing card is drawn. Find the probability of (i) a six and an ace, (ii) a six or an ace, (iii) a six but not an ace.

3. Two dice are thrown. Find the chance that (i) they are both fours, (ii) neither is a four, (iii) only one is a four.

4. The probability of three cars calling for petrol are $\frac{1}{4}$, $\frac{3}{5}$, and $\frac{2}{3}$ respectively. What is the probability that (i) none calls, (ii) at least one calls.

5. A room has ten lights controlled by ten switches. If four of the bulbs are broken, find the chance that there will be no light if two are switched on.

6. A car park has 5 Jaguars, 4 Rovers and 3 Rolls Royces parked. Find the chance that the first to leave is a Jaguar or a Rolls Royce, and the second is a Jaguar or a Rover.

7. Find the probability in Question 6 of one, and only one, Rolls Royce leaving in the first three.

8. Find the probability in Question 6 of a Jaguar followed by a Rover, followed by a Rolls Royce, and so on, until only one car, a Jaguar is left.

31.7

In this section we shall consider that if a special feature happens in an event it will be a success, and if it does not, it is a failure. The probabilities of success and failure are p and q respectively. If the event takes place n times independently, then the expected probabilities of the various combinations of success and failure are given by the terms of the binomial expansion $(q+p)^n$.

$$(q+p)^n = \qquad q^n \qquad \text{(all failures)}$$
$$+ nq^{n-1} \cdot p \qquad \text{(all but one are failures)}$$
$$+ \frac{n.(n-1)}{1.2} \cdot q^{n-2} \cdot p^2 \qquad \text{(all but two are failures, i.e. two successes)}$$
$$\cdots\cdots\cdots$$
$$\cdots\cdots\cdots$$
$$+ p^n \qquad \text{(all successes)}$$

The sum of the probabilities is $(q+p)^n$, but $q = 1-p$ giving $(1-p+p)^n = 1$, which is the total probability.

We have taken the probability that a Ford is the first car to leave our park as $\frac{1}{5}$. What are the expected probabilities of a Ford being first 0, 1, 2, 3, 4, or 5 times in a five-day week (once per day)?

$$p = \tfrac{1}{5}; \quad q = \tfrac{4}{5}; \quad n = 5$$

$$
\begin{aligned}
(q+p)^n = (\tfrac{4}{5}+\tfrac{1}{5})^5 & \\
= \quad (\tfrac{4}{5})^5 & \qquad \text{Probability of no Fords first} \\
+ \; 5(\tfrac{4}{5})^4.(\tfrac{1}{5}) & \qquad ,, \qquad ,, \text{ one Ford } ,, \\
+ 10(\tfrac{4}{5})^3.(\tfrac{1}{5})^2 & \qquad ,, \qquad ,, \text{ two Fords } ,, \\
+ 10(\tfrac{4}{5})^2.(\tfrac{1}{5})^3 & \qquad ,, \qquad ,, \text{ three } ,, \quad ,, \\
+ \; 5(\tfrac{4}{5}) \;.(\tfrac{1}{5})^4 & \qquad ,, \qquad ,, \text{ four } ,, \quad ,, \\
+ \quad (\tfrac{1}{5})^5 & \qquad ,, \qquad ,, \text{ five } ,, \quad ,,
\end{aligned}
$$

If the cycle of n events is repeated N times, the expected number of times that the various combinations, of success and failure, occur are given by the expansion of $N(q+p)^n$.

If we carry out our weekly observations for 125 weeks, the number of weeks that Fords are expected to be first twice during the five days is

$$125 . 10 . \left(\frac{4}{5}\right)^3 \left(\frac{1}{5}\right)^2 = \frac{128}{5} = 25\tfrac{3}{5}$$

This would be rounded off to 26 weeks.

We can use the theorem of total probabilities to find in how many weeks Fords are expected to be first at least three times, i.e. 3, 4, or 5.

$$125 . \frac{1}{5^5}(10 . 16 + 5 . 4 + 1) = \frac{181}{25} = 7\tfrac{6}{25}$$

If we require the successes rather than failures we can expand $(p+q)^n$ as far as is necessary. In experiments special precautions are taken to keep p constant. If a card is drawn from a pack it is returned before the next draw is made, keeping p as $\frac{1}{52}$. Also the selection may be taken from a large sample.

Exercise 31b

1. What is the chance of drawing one of the 13 clubs from a pack of 52 playing cards if one draw is made? Four cards are drawn from a pack, each being returned before the next is drawn. This is repeated 400 times. How many times can we expect at least two clubs out of four?

2. What is the chance of a single coin coming down heads? A group of 8 coins are tossed 256 times. How many times can we expect a majority of heads out of the eight?

3. Three dice are thrown and the number of ones or sixes recorded. This is done 108 times. Find the theoretical or expected frequency that 0, 1, 2, and 3 successes are recorded.

4. The chance of catching measles whilst under 15 years old is $\frac{4}{5}$. What is the chance that in a group of 10 pupils of 15 years old not more than two will have had measles? (Leave the denominator in powers of 5.)

5. In a large sample of electric light bulbs 1 in 25 is sub-standard. If 4 bulbs are selected, calculate the probability that (i) all, (ii) none, will be sub-standard.

6. England's chance of winning a test match is $\frac{2}{3}$. Find their chance of regaining the Ashes during the five-match series, assuming that brighter cricket leads to a result in each match and there are no ties.

31.8

For the binomial distribution $N(q+p)^n$ it can be shown that:

$$\text{Mean} = n.p$$

$$\text{Standard deviation} = \sqrt{n.p.q}.$$

For the experiment over 125 weeks we had $n = 5$ days, $p = \frac{1}{5}$ for a Ford, and $q = \frac{4}{5}$.

$$\text{Mean} = \frac{5}{5} = 1 \text{ Ford first in each week.}$$

$$\text{Standard deviation} = \sqrt{5 \cdot \frac{1}{5} \cdot \frac{4}{5}} = \frac{2\sqrt{5}}{5}$$

Exercise 31c. (Leave as surds where they occur.) Find the mean and standard deviation of Questions 1–6 of the previous exercise.

31.9

If a large population has a proportion p in which a special feature occurs we must consider what to expect if we take a sample of that population. Suppose it is known that one in five of the cars on our roads is made by Fords. If we note down the names of a hundred cars that pass us we should hardly expect twenty to be Fords. Within what limits can the number of Fords lie for us to accept the proportion as correct?

A sample of n is taken and x of these have the special feature which is in the proportion p for the whole population. If we took repeated samples of n things the mean values of x for all the samples would be $n.p$, and the standard deviation $\sqrt{n.p.q}$, as stated in the previous section.

$$\therefore p = \frac{\text{mean value of } x}{n}$$

and the standard deviation of this proportion would be

$$\frac{\sqrt{npq}}{n} = \sqrt{\frac{pq}{n}}$$

This may be referred to as the standard error of the proportion.

For our single sample of n the special feature occurs x times, therefore the experimental proportion is $\frac{x}{n}$. If this varies from p by an amount less than twice the standard deviation $\sqrt{\frac{pq}{n}}$, the result can be accepted. The difference is due to the small sample. If it varies by between twice and three times the standard deviation we can be reasonably sure that the result is significant, This is called the 5 % level, which is the greatest chance that it will happen. For a variation of more than three times the standard deviation we can be confident that it is significant (1 % level). If the sample was random a significant result means that there is bias which has made the proportion p incorrect.

Taking our sample of 100 cars:

$$n = 100, \quad p = \tfrac{1}{5} \quad \text{and} \quad q = \tfrac{4}{5}$$

$$\text{Standard deviation} = \sqrt{\frac{p.q}{n}} = \sqrt{\frac{1}{5}.\frac{4}{5}.\frac{1}{100}} = \frac{1}{25} = 0.04$$

For 5 % level of significance $p \pm 2$ s.d. $= 0.20 \pm 0.08 = 0.12$, or 0.28. This means that for any total between 13 or 27 Fords out of the 100 there is no significance.

For 1 % level of significance $p \pm 3$ s.d. $= 0.20 \pm 0.12 = 0.08$, or 0.32.

This means that for a total between 8 and 12, or 28 and 32, we can be reasonably sure of significance; 7 and below or 33 and above is significant. $\left(\text{Suppose that we counted 15 Fords:} \frac{x}{n} = \frac{15}{100} = 0.15.\right)$

Notice that if we increase n to 400 the standard deviation is

$$\sqrt{\frac{1}{5}.\frac{4}{5}.\frac{1}{400}} = \frac{1}{50} = 0.02$$

For the 5 % level of significance $p \pm 2$ s.d. $= 0.16$, or 0.24. These values are nearer to p because the larger value of n should lead to less sampling error. In reverse, if we know the proportion of a sample the proportion of the whole population will be likely to be within three times the standard deviation from the proportion.

Exercise 31d

1. Theory suggests that 1 in 4 families own a car. If 114 families own cars out of 400 questioned, is there evidence of significance?

2. A die is thrown 120 times. Is there evidence of bias if the number of ones are (*a*) 14, (*b*) 29?

3. In a sample of 500 valves 50 were sub-standard. Use the standard error to show that the full consignment will have less than 14% of sub-standard valves.

4. A coin was tossed 200 times and came down heads 124 times. What is the standard error of the proportion?
   Is there evidence of bias?

5. A random sample of 500 eggs was found to have 56 bad ones. Deduce as a percentage the limits within which the number of bad eggs for the whole consignment is likely to lie.

6. It is thought that 5% of cars are convertibles. If 1000 are taken as a sample is there evidence of bias if (*a*) 59, (*b*) 35, (*c*) 72 are convertibles?

31.10

If a sample of n has a mean $m(\bar{x})$, and a standard deviation s we can obtain an estimate of the limits within which the population mean will fall. For the mean m it can be shown that the standard deviation (or standard error) of the mean is $\dfrac{s}{\sqrt{n}}$ (s being an estimate of the population standard deviation σ).

By the 5% level of significance we can be reasonably sure that the population mean (μ) lies within the range $m \pm \dfrac{2s}{\sqrt{n}}$. At the 1% level we can be confident that the mean lies within the range $m \pm \dfrac{3s}{\sqrt{n}}$.

The multiple of the standard error for the 5% and 1% levels varies with the size of n, being 1·96 and 2·58 for an infinite population. The values 2 s.e. (5%) and 3 s.e. (1%) are the usual figures for small samples of more than 30 and 12 respectively.

If a sample of 64 objects has a mean 100 cm and a standard deviation 4 cm, the standard error is given by

$$\frac{s}{\sqrt{n}} = \frac{4}{\sqrt{64}} = 0.5 \text{ cm}$$

The mean of the whole consignment is probably between 99 cm and 101 cm, and almost certainly between 98·5 cm and 101·5 cm.

31.11

If we take r samples each of n observations we can estimate the mean
(μ) and standard deviation (σ) of the whole population. The first
sample will have mean m_1, the second sample m_2, and so on up to m_r.
The mean of the means

$$\bar{m} = \frac{m_1 + m_2 + \ldots + m_r}{r}$$

and, remembering that each sample had an equal number of observations, it can be shown that $\mu = \bar{m}$.

The standard deviation (s_m) for the means $m_1, m_2, m_3, \ldots m_r$ is then
found. The standard deviation of the population $\sigma = s_m \cdot \sqrt{n}$. The
population standard deviation is the standard deviation for individual observations and we should expect it to be larger than the
standard deviation of the means (s_m) as the individual observations
will be more scattered than the means.

The heights of 30 sample groups each of 16 Christmas trees were
measured and the group means calculated. If the mean of the means
was 40 cm, with standard deviation 3 cm, within what limits would
we expect the group means to lie and what is the standard deviation
for individual trees? $r = 30$, $n = 16$, $\bar{m} = 40$ cm, $s_m = 3$ cm.

The means are likely to be within three times the standard deviation (1% level), i.e. 40 cm $\pm$ 9 cm. They lie between 31 cm and 49 cm.

The population or individual mean (μ) will be 40 cm and the
standard deviation (σ) $= s_m \cdot \sqrt{n} = 3\sqrt{16} = 12$ cm.

Notice that the individual members will almost certainly lie within
the range 40 ± 36, which is a much greater scatter than the means.

Exercise 31e. Find the limits within which you can be confident
(1% level) the population mean lies for the following samples:

1. 100 objects, mean 21·6 cm standard deviation 4·2 cm.
2. 256 objects, mean 280, standard deviation 20.
3. 64 objects, mean 280, standard deviation 20.
4. 16 objects, mean 280, standard deviation 20.

Estimate the range (1% level) within which the means are likely
to lie, also the individual standard deviation for the following groups
of samples:

5. Groups of 64 trees, with mean of means 72 cm and standard
   deviation 5 cm.
6. Groups of 256 trees, with mean of means 72 cm and standard
   deviation 5 cm.
7. 25 groups of 49 trees, with mean of means 48 cm and standard
   deviation 1 cm.

8. The standard deviation and mean of a sample are 4 cm and 100 cm. How large was the sample to ensure that it was unlikely (1 % level) that the mean varied from the population mean by more than 1 % of itself.

9. Find the mean, standard deviation, and standard error of the mean for the sample integers 1 to 9.

ANSWERS

1a. 1. $1 + 4x + 6x^2 + 4x^3 + x^4$
  2. $1 + 6x + 15x^2 + 20x^3 + 15x^4 + 6x^5 + x^6$
  3. $1 + 5x + 10x^2 + 10x^3 + 5x^4 + x^5$
  4. $1 + 8x + 28x^2 + 56x^3 + 70x^4 + 56x^5 + 28x^6 + 8x^7 + x^8$

1b. 1. $1 - 9x + 36x^2 - 84x^3 + 126x^4 - 126x^5 + 84x^6 - 36x^7 + 9x^8 - x^9$
  2. $1 + 12y + 60y^2 + 160y^3 + 240y^4 + 192y^5 + 64y^6$
  3. $1 - 5x + 10x^2 - 10x^3 + 5x^4 - x^5$
  4. $1 + 12y + 54y^2 + 108y^3 + 81y^4$
  5. $1 - 12x + 60x^2 - 160x^3 + 240x^4 - 192x^5 + 64x^6$
  6. $1 - 12x + 54x^2 - 108x^3 + 81x^4$

1c. 1. $0 \cdot 886$ 3. $1 \cdot 08286$
  2. $1 \cdot 0510$ 4. $0 \cdot 886$

1d. 1. $a^5 + 5a^4 x + 10a^3 x^2 + 10a^2 x^3 + 5ax^4 + x^5$
  2. $a^8 + 8a^7 x + 28a^6 x^2 + 56a^5 x^3 + 70a^4 x^4 + 56a^3 x^5 + 28a^2 x^6 + \\ + 8ax^7 + x^8$
  3. $a^5 - 5a^4 x + 10a^3 x^2 - 10a^2 x^3 + 5ax^4 - x^5$
  4. $a^4 + 12a^3 y + 54a^2 y^2 + 108ay^3 + 81y^4$
  5. $64a^6 + 192a^5 x + 240a^4 x^2 + 160a^3 x^3 + 60a^2 x^4 + 12ax^5 + x^6$
  6. $16b^4 - 96b^3 y + 216b^2 y^2 - 216by^3 + 81y^4$

1e. 1. $180a^8 b^2$ 2. $560a^6 x^4$ 3. 240

  4. $40y$ 5. $\dfrac{70x^4}{y^4}$ 6. $5\frac{11}{16}\left(\frac{91}{16}\right)$

2a. 1. $4x^3$ 2. 1 3. $5x^4$ 4. $8x^7$

2b. 1. $\frac{1}{3}.x^{-2/3}$ 2. $-2x^{-3}$ 3. $\frac{1}{4}.x^{-3/4}$
  4. $-3.x^{-4}$ 5. $-\frac{1}{2}.x^{-3/2}$ 6. $-\frac{1}{3}.x^{-4/3}$

2c. 1. $9x^8$ 2. $27x^{26}$ 3. $-7x^{-8}$ 4. $-9x^{-10}$
  5. $\frac{1}{5}.x^{-4/5}$ 6. $\frac{1}{6}.x^{-5/6}$ 7. $\frac{2}{5}.x^{-3/5}$ 8. $\frac{3}{7}.x^{-4/7}$
  9. $-\frac{1}{5}.x^{-6/5}$ 10. $-\frac{2}{5}.x^{-7/5}$ 11. $-\frac{3}{7}.x^{-10/7}$ 12. $0 \cdot 6.x^{-0 \cdot 4}$

3a.   1. $\dfrac{\pi}{2}$ 2. $\dfrac{\pi}{4}$ 3. $\dfrac{\pi}{3}$ 4. $\dfrac{3\pi}{4}$

      5. $\dfrac{5\pi}{3}$ 6. $\dfrac{2\pi}{3}$ 7. $\dfrac{9\pi}{4}$ 8. $\dfrac{\pi}{18}$

      9. $45°$ 10. $30°$ 11. $120°$ 12. $540°$

    13. $18°$ 14. $72°$ 15. $80°$ 16. **310°**

3b.   1. $\sin 60°, \sin 15°, \sin 45°$

      2. $\tan 75°, \tan 30°, \tan 0°$

      3. $\operatorname{cosec} 70°, \operatorname{cosec} 35°, \operatorname{cosec} 10°$

      4. $\sec 69°, \sec 45°, \sec 14°$

3d.   1. $\sqrt{2}$ 2. 1 3. $\sqrt{3}$ 4. $\dfrac{2}{\sqrt{3}}$

      5. 2 6. $\dfrac{2}{\sqrt{3}}$ 7. $\dfrac{1}{\sqrt{3}}$ 8. $\sqrt{2}$

      9. 2

4b.   1. $12, 12$ 2. $-\frac{1}{4}, -\frac{1}{4}$ 3. $+\frac{1}{4}, -\frac{1}{4}$

4c.   1. $\frac{3}{2}, 2$ 2. $\frac{1}{2}, -\frac{2}{3}$ 3. $-\frac{2}{3}, 1$

      4. $\frac{3}{2}, -\frac{1}{2}$ 5. $\frac{3}{2}, \frac{5}{2}$ 6. $2, -3$

4d.   1. 19 2. $-\frac{3}{2}$ or -1

      3. $(-1, 8)$ and $(1, 0)$ 4. 1

      5. $y = 3x - 7$ 6. $2y = x - 8$, no

5a.   1. $4x^3 + 3x^2$ 2. $1 - \frac{1}{2} \cdot x^{-1/2}$ 3. $4x^3 - \dfrac{4}{x^5}$

      4. $12x$ 5. $21x^2 - \dfrac{2}{x^2}$ 6. $x^2 + \dfrac{1}{x^3}$

      7. $12x + 5$ 8. $2x - \dfrac{12}{x^4}$ 9. $3x^{-4/7}$

    10. $+5$ at $x = 4$; -5 at $x = -1$

    11. $+6$ at $x = 2$; -2 at $x = 0$; $+3$ at $x = -1$

    12. -4 at $x = 2$; $+4$ at $x = -2$

5b.   1. $6x^5$ 2. $36x + 27$

      3. $12x^2 + 8x + 1$ 4. $18x^2 + 14x + 2$

      5. $-18x^2 + 10x + 14$ 6. $1 - \dfrac{2}{x^2}$

      7. $\frac{15}{2} x^{3/2}$ 8. 20 at $x = 2$; -4 at $x = -2$

                                     5 at $x = -3$

5c.   1. $72x^7$ 2. $702x^{25}$ 3. $56x^{-9}$ 4. $90x^{-11}$

     5. $-\frac{4}{25} \cdot x^{-9/5}$ 6. $-\frac{5}{36}x^{-11/6}$ 7. $-\frac{6}{25}x^{-8/5}$ 8. $-\frac{12}{49}x^{-11/7}$

     9. $\frac{6}{25}x^{-11/5}$ 10. $\frac{14}{25}x^{-12/5}$ 11. $\frac{30}{49}x^{-17/7}$ 12. $-0 \cdot 24x^{-1 \cdot 4}$

     13. 0

6a.   1. 380   2. 24   3. 5040

     4. 2520   5. 720   6. 15,120

6b.   1. 720; 120   2. 576; 36   3. (a) 24; (b) 18

     4. 114   5. 5760   6. 100; 32

     7. 120   8. 24

6c.   1. 120; 48   2. 216   3. 30; 6   4. 1500

     5. 60; 24; 33; 12

     6. (a) 180; (b) 12; (c) 19,958,400

     7. (a) 45,360; (b) 60

     8. 62

6d.   1. 10   2. 165   3. 74

     4. 200   5. 350; 170   6. 220; 110

6e.   1. 1023   2. (a) 1 in 12; (b) 1 in 6; (c) 1 in 4

     3. 1 in 2   4. 1 in 2   5. 31

     6. 105   7. 2 in 5, 4 in 11   8. 4 in 33

7a.   1. $30°$ 2. $359°$ 3. $267°$ 4. $65°$

     5. $190°$ 6. $203°$ 7. $90°$ 8. $190°$

     9. $180°$ 10. $41°$ 11. $319°$ 12. $320°$

7b.   1. $70°$ 2. $10°$ 3. $80°$ 4. $40°$

     5. $36°$ 6. $85°$ 7. $70°$ 8. $30°$

     9. $30°$ 10. $60°$ 11. $10°$ 12. $30°$

7c.   1. $0 \cdot 9848$ 2. $-0 \cdot 8660$ 3. $-0 \cdot 8660$ 4. $0 \cdot 9455$

     5. $-0 \cdot 9848$ 6. $0 \cdot 6428$ 7. $-1 \cdot 0$ 8. $-0 \cdot 7071$

     9. $-0 \cdot 9397$

7d.   1. $-0 \cdot 3420$ 2. $-0 \cdot 9397$ 3. $0 \cdot 7660$ 4. $0 \cdot 9848$

     5. $-0 \cdot 5000$ 6. 0   7. $0 \cdot 7071$ 8. $0 \cdot 8660$

     9. $0 \cdot 9744$

7e.   1. $-5 \cdot 671$ 2. $0 \cdot 3640$ 3. $-1 \cdot 1918$ 4. 0

     5. $5 \cdot 671$ 6. $11 \cdot 43$ 7. $1 \cdot 1547$ 8. $1 \cdot 1547$

     9. $2 \cdot 747$

7f. 1. $-\frac{4}{3}$ 2. $-\frac{1}{2}$ 3. $\frac{1}{\sqrt{2}}, \sqrt{2}$ 4. $-2\frac{2}{5}$

5. $54°\,28',\ 234°\,28',\ 0{\cdot}8138,\ -0{\cdot}8138$

6. $-\frac{5}{13}$ 7. $-\frac{1}{2}$

8. $30°,\ 150°,\ 0{\cdot}8660,\ -0{\cdot}8660$

9. $-\frac{15}{17},\ -\frac{15}{8}$

7g. 1. $30°,\ 131°\,48',\ 150°,\ 228°\,12'$

2. $19°\,28',\ 160°\,32',\ 210°,\ 330°$

3. $60°,\ 90°,\ 270°,\ 300°$

4. $108°\,26',\ 116°\,34',\ 288°\,26',\ 296°\,34'$

5. $63°\,26',\ 71°\,34',\ 243°\,26',\ 251°\,34'$

6. $45°,\ 135°,\ 225°,\ 315°$

7. $18°\,26',\ 198°\,26'$

8. $7°\,1',\ 90°,\ 97°\,1',\ 187°\,1',\ 270°,\ 277°\,1'$

9. $60°,\ 70°\,32',\ 289°\,28',\ 300°$

10. $0°,\ 360°$

11. $35°\,47',\ 125°\,47',\ 215°\,47',\ 305°\,47'$

12. $30°,\ 150°,\ 270°$

8a. 1. 10   2. $5{\cdot}66$ 3. 13

4. $4{\cdot}12$ 5. $2{\cdot}83$ 6. $5{\cdot}10$

8b. 1. $(4, 5)$ 2. $(8, 9)$ 3. $(-1, 0)$

4. $(3\frac{1}{2}, 4\frac{1}{2})$ 5. $(3, 4)$ 6. $(-\frac{9}{7}, -\frac{1}{7})$

7. $(\frac{25}{2}, -\frac{31}{2})$ 8. $(\frac{1}{2}, \frac{5}{2})$

8c. 1. $-\frac{4}{5}$ 2. $-\frac{3}{2}$ 3. $12°\,32'$

4. $11°\,53'$ 5. $130°\,36'$ 6. $90°$

8d. 1. $5\frac{1}{2}$ sq. units   2. 7 sq. units   3. $13\frac{1}{2}$ sq. units

4. $(-9, 0)$, or $(\frac{53}{3}, 0)$

5. zero, straight line

6. $(-2, 1); (-6, 9); (6, -5)$

9a. 1. $24x^{23}$ 2. $\frac{3}{5}.x^{-2/5}$ 3. $\dfrac{-14{,}406}{x^3}$

4. $\dfrac{5x^4}{7y^6}$ 5. $\dfrac{1+2x}{10y^4}$ 6. $\dfrac{3x^2+2x+3}{2y}$

7. $\dfrac{1}{7y^6}$ 8. $\dfrac{1}{6y^2}$

9b. 1. $3(x+7)^2$ 2. $10(2x+1)^4$

3. $24x(2x^2+7)^5$ 4. $\dfrac{x}{(x^2+1)^{1/2}}$

5. $\dfrac{2x}{3(x^2+1)^{2/3}}$ 6. $\dfrac{(10x^4+3)}{4(2x^5+3x)^{3/4}}$

7. $\dfrac{-4x}{(x^2+1)^3}$ 8. $\dfrac{-6x}{(x^2+1)^4}$

9. $\dfrac{-4(10x^4+3)}{(2x^5+3x)^5}$ 10. $72x+63$

11. $6x^5+12x^3+8x-2$ 12. $2+\dfrac{x}{(x^2+1)^{1/2}}$

9c. 1–12. See 9b.

13. $4(1+x^2)(5x^2+3x+1)$

14. $2(2+x)(3-x^2)(3-4x-3x^2)$

10a. 1. $B = 44° 26'$; $C = 105° 34'$; $c = 9·63$ cm.

2. $c = 2·49$ cm.; $B = 135° 34'$; $C = 14° 26'$

3. $B = 20° 29'$; $C = 129° 31'$; $c = 7·72$ cm.

4. $a = 4·71$ m ; $b = 6·34$ m ; $C = 80°$

5. $a = 2·59$ m ; $A = 15° 23'$; $C = 114° 37'$

6. $a = 3·11$ m ; $c = 2·42$ m ; $C = 30°$

7. $a = 8·86$ m ; $A = 64° 37'$; $C = 65° 23'$

8. $b = 11·9$ m ; $B = 65° 12'$; $C = 44° 48'$

10b. 1. $A = 17° 9'$; $B = 45° 2'$ $C = 117° 49'$

2. $A = 23° 26'$; $B = 72° 40'$; $C = 83° 54'$

3. $a = 4·58$ cm.; $B = 70° 53'$; $C = 49° 7'$

4. $a = 6·39$ cm.; $B = 147° 37'$; $C = 12° 23'$

5. $A = 130° 33'$; $B = 22° 19'$; $C = 27° 8'$

6. $c = 14·5$ cm.; $A = 13° 59'$; $B = 16° 1'$

7. $c = 4·00$ cm.; $A = 61° 1'$; $B = 88° 59'$

8. $A = 38° 53'$; $B = 40° 3'$; $C = 101° 4'$

10c. 1. 6 cm^2 2. $70·8$ cm^2

3. $6·87$ cm^2 4. $4·43$ cm^2

5. $20·7$ cm^2 6. $149° 16'$, or $30° 44'$

7. $26·8$ cm^2 8. 84 cm^2

11a.   1. 119   2. 13 m/s, 2 m/s^2
    3. 0, or $\frac{4}{3}$ s   4. $1\frac{2}{5}$ m/s
    5. 408 m/s^2 6. ± 0.03 m^2
    7. 30 m, 17 m/s, 12 m/s^2
    8. 0·0063 m^2; 0·0063 m

11b.   1. $(-\frac{3}{5}, -\frac{14}{5})$, min.   2. $\frac{16}{5}$, max.
    3. -27, min.; -26, max.
    4. 9, max; 5, min.

    5. $\left(-\dfrac{1}{\sqrt{3}}, -2\sqrt{3}\right)$, max.; $\left(+\dfrac{1}{\sqrt{3}}, +2\sqrt{3}\right)$, min.

    6. $(\frac{1}{2}, \frac{3}{2})$, min.
    7. $p = -\frac{3}{2}, q = -6$; $(-1, \frac{15}{2})$, max.; $(+2, -6)$, min.
    8. $p = -1, q = \frac{3}{2}$; $y = 0$, min.; $y = \frac{1}{2}$, max.
    9. $y = -3$, max.
    11. $\frac{1}{4}$ m^2 12. $h = \sqrt{2} . r$

12a.   1. $y^{5/2}, p^{1/6}$ 2. $36y^2 . x^{12}$
    3. $x^{-11/6}$ 4. $9x^{-5/3}$
    5. (a) 8; (b) 8; (c) 16   6. 12
    7. (a) $\frac{5}{3}$; (b) $\frac{3}{2}$ 8. $3x^{-13/6}$

12b.   2. $\frac{5}{3}$ 3. $-0.7213, \bar{1}.2787$
    4. (a) $6p$; (b) $2(q-1)$
    5. (a) $\log_{10}2 + \log_{10}3$; (b) $2\log_{10}3 - \log_{10}2$; $\log_{10}3 = 0.47712$
    6. $q = 500p^3$ 7. 2
    8. (a) $\log p^3$; (b) $\frac{5}{2}$ 9. $p = 2\frac{1}{2}, q = 250$
    10. 2·03   11. $y = 0.188$
    12. 7·34

12c.   1. (a) $\dfrac{\sqrt{2}}{2}$; (b) $\dfrac{2\sqrt{3}}{3}$; (c) $2\sqrt{3}$; (d) $\dfrac{\sqrt{14}}{7}$

    2. (a) $\dfrac{\sqrt{3}-1}{2}$; (b) $\dfrac{(1+\sqrt{3})}{2}$; (c) $2+\sqrt{3}$

    3. $x-1$
    4. $b = -6, c = 11$

    5. (a) $\dfrac{-(11+6\sqrt{2})}{7}$; (b) $\dfrac{6+3\sqrt{3}-2\sqrt{2}-\sqrt{6}}{7}$

    6. $a = -7, b = -27$

7. $21 - 2\sqrt{3}$

9. (a) $4\sqrt{3}$; (b) $6\sqrt{2}$; (c) $10\sqrt{3}$

10. $\dfrac{-(6\sqrt{2} + 12\sqrt{6} + 20\sqrt{3} + 10)}{7}$

11. (a) $5\sqrt{2}$; (b) $3(\sqrt{3} + \sqrt{5})$; (c) $3\sqrt{2} + 2\sqrt{3}$

12. $a = 5$, $b = 6$, $(3x - 1)(x + 1)$

13a. 1. $\dfrac{x^5}{5} + c$ 2. $\dfrac{x^8}{8} + c$

3. $\dfrac{x^{10}}{10} + c$ 4. $\dfrac{x^{14}}{14} + c$

5. $\dfrac{x^{22}}{22} + c$ 6. $\dfrac{x^2}{2} + \dfrac{x^9}{9} + c$

7. $\dfrac{x^{11}}{11} + \dfrac{x^4}{4} + c$ 8. $\dfrac{x^{13}}{13} - \dfrac{x^{12}}{12} + c$

13b. 1. $-\dfrac{1}{x} + c$ 2. $-\dfrac{1}{3x^3} + c$ 3. $-\dfrac{1}{6x^6} + c$

4. $\dfrac{3x^{4/3}}{4} + c$ 5. $\dfrac{4x^{5/4}}{5} + c$ 6. $\dfrac{7x^{8/7}}{8} + c$

7. $\dfrac{2x^{3/2}}{3} + \dfrac{1}{x} + c$ 8. $\dfrac{x^7}{7} + \dfrac{1}{5x^5} + c$ 9. $\dfrac{x^7}{7} - \dfrac{1}{5x^5} + c$

13c. 1. $-4x^{-1/4} + c$ 2. $5x^{1/5} + c$ 3. $\dfrac{5x^{9/5}}{9} + c$

4. $-\dfrac{2}{3x^{3/2}} + c$ 5. $5x + c$

6. $7x^3 - 2x^2 + 3x + c$ 7. $x^5 - 6x + c$ 8. $-\dfrac{2}{3x^{3/2}} + c$

9. $2x^3 + 6x - \dfrac{6}{x} + c$ 10. $\dfrac{21x^{22}}{22} + c$ 11. $\dfrac{5x^{4/5}}{4} + c$

12. $\dfrac{17x^{19}}{19} + c$

13d. 1. $\tfrac{1}{6}.(3 + x)^5 + c$ 2. $-(2 - x)^3 + c$ 3. $\tfrac{1}{25}.(5x + 4)^5 + c$

4. $\tfrac{1}{8}.(x^2 + 1)^4 + c$ 5. $-\dfrac{1}{(2 + x)} + c$ 6. $-\tfrac{2}{3}.(3 - x)^6 + c$

7. $-\dfrac{7}{4(1 + x^2)^2} + c$ 8. $-\dfrac{1}{3(3 + 5x^3)} + c$ 9. $\tfrac{2}{3}.(3 + x)^{3/2} + c$

10. $-\tfrac{2}{9}.(1 - x^3)^{3/2} + c$ 11. $(1 + 2x)^{1/2} + c$ 12. $\tfrac{1}{4}.(x^3 + x^2)^4 + c$

14a. 1. 0·1736   2. −0·8660   3. 0·8660   4. 0·7660

5. −0·6428   6. −0·3420   7. 0·3420   8. −0·9848

9. −0·9848   10. −2·0000   11. −1·5557   12. 1·5557

13. $\cos A$ 14. $\sin A$ 15. $-\cos A$ 16. $-\sin A$

17. $\cos A$ 18. $\sin A$ 19. $-\cos A$ 20. $-\sin A$

21. $-\sin A$ 22. $\operatorname{cosec} A$ 23. $-\operatorname{cosec} A$ 24. $-\operatorname{cosec} A$

14b. 1. −2·747   2. 0·3640   3. −5·671   4. −1·1918

5. −0·8391   6. −0·3640   7. $\tan A$ 8. $\tan A$

9. $-\cot A$ 10. $-\cot A$ 11. $-\cot A$ 12. $\tan A$

14c. 1. $4\cos^3\theta - 3\cos\theta$

2. (i) $2\cos^2 2\theta - 1$; (ii) $8\cos^4\theta - 8\cos^2\theta + 1$

3. $\dfrac{3\tan\theta - \tan^3\theta}{1 - 3\tan^2\theta}$

14d.   (roots taken positive).

1. $\dfrac{\sqrt{2}(\sqrt{3}-1)}{4}$ 2. $\sqrt{2}(\sqrt{3}-1)$ 3. $2-\sqrt{3}$

4. $\dfrac{\sqrt{2}(\sqrt{3}+1)}{4}$ 5. $\dfrac{\sqrt{2}(\sqrt{3}-1)}{4}$ 6. $\dfrac{\sqrt{2}(\sqrt{3}-1)}{4}$

7. -1 8. $-\frac{1}{2}$ 9. $2+\sqrt{3}$

10. $-(2+\sqrt{3})$ 11. $\dfrac{-\sqrt{2}(\sqrt{3}-1)}{4}$ 12. $-\sqrt{2}(\sqrt{3}+1)$

14e. 1. 37 m   2. N. 82° 9′ W.   3. 196 m

4. 1350 m

5. 12·4 cm.

6. (i) 71° 37′; (ii) 64° 50′; (iii) 95° 42′

15a. 1. $\frac{1}{6}$ sq. units   2. $291\frac{3}{4}$ 3. $1\frac{1}{3}$ sq units
                                                          (below axis)

4. $1\frac{1}{3}$ sq units.   5. 1 sq. unit   6. $\frac{1}{6}$ sq. units

7. $12\frac{4}{5}$ 8. $1\frac{1}{2}$ 9. triangle

10. $\frac{1}{20}$ sq. units

15b. 1. (i) $\dfrac{5\pi}{2}$ cu. units; (ii) $\dfrac{\sqrt{5}\pi}{5}$ cu units

2. 1800π cu. units, cone

3. $\dfrac{128\pi}{7}$ cu. units   4. $\frac{1}{3}\pi r^2 h$

5. 9π cu. units   6. $\frac{4}{3}\pi a^3$

7. $\dfrac{\pi}{30}$ cu. units   8. $\dfrac{\pi}{5}$ cu. units

16a. 1. $2x-y-3=0$ 2. $3x-2y-8=0$
   3. $x+y+2=0$ 4. $x-y=0$
   5. $x+y+5=0$ 6. $5x+6y+8=0$
   7. $y=1$ 8. $y=2$

16b. 1. $2x+3y-7=0$ 2. $3x-2y-4=0$
   3. $x+2y-3=0$ 4. (i) $y=1$; (ii) $x=2$; (iii) $2y=x$
   5. $5x-3y+15=0$ 6. $x-2y-4=0,\ m=\frac{1}{2}$

16c. 1. (i) $2=\dfrac{\sqrt{3}}{2}x+\frac{1}{2}y$ (ii) $\sqrt{3}x+y-4=0$

2. (i) $4=\dfrac{-1}{\sqrt{2}}x+\dfrac{1}{\sqrt{2}}y$ (ii) $x-y+4\sqrt{2}=0$

3. (i) $3=-1.x$ (ii) $x+3=0$

4. (i) $1=\frac{1}{2}x-\dfrac{\sqrt{3}}{2}y$ (ii) $x-\sqrt{3}y-2=0$

5. $3=\frac{3}{5}x+\frac{4}{5}y$ 6. $4\sqrt{2}=\dfrac{1}{\sqrt{2}}x+\dfrac{1}{\sqrt{2}}y$

7. $\frac{7}{2}=-\dfrac{\sqrt{3}}{2}x+\frac{1}{2}y$ 8. $2=-\frac{8}{17}x+\frac{15}{17}y$

16d.   (roots taken positive).

1. $\dfrac{3\sqrt{3}-5}{2}$ 2. $\dfrac{-3(\sqrt{3}+2)}{2}$

3. $-\dfrac{(3\sqrt{2}+10)}{2}$ 4. 5

5. -1 6. -20

7. $-(5-\sqrt{3})$ 8. 28

(When p is negative, point is on same side as origin.)

17a. 1. 39,210   2. 988   3. -24
   4. 1, 3, 5; 144   5. $\frac{4}{5}$ 7. $10\frac{1}{2}$, 6, $1\frac{1}{2}$
   8. 2

P

17b. 1. $3\frac{3}{8}$, $13\frac{3}{16}$ 2. $25\cdot97$ 3. 9

4. (i) -125; (ii) 64   5. 6

6. (a) $\frac{1}{2}$; (b) 1; (c) $511\frac{1}{2}$

7. (i) $6\frac{1}{3}$ cm., $8\frac{2}{3}$ cm.; (ii) 6 cm., 9 cm.   8. 3

17c. 1. $1 - 2x + 3x^2 - 4x^3 + 5x^4$

2. $1 + \dfrac{x}{2} - \dfrac{x^2}{8} + \dfrac{x^3}{16} - \dfrac{5x^4}{128} + \dfrac{7x^5}{256}$

3. $1 - \dfrac{x}{3} + \dfrac{2x^2}{9} - \dfrac{14x^3}{81} + \dfrac{35x^4}{243}$

4. $1 + 4x + 10x^2 + 20x^3 + 35x^4$

5. $\dfrac{1}{\sqrt{a}} - \dfrac{x}{2\sqrt{a^3}} + \dfrac{3x^2}{8\sqrt{a^5}} - \dfrac{5x^3}{16\sqrt{a^7}}$

6. $1 + x + x^2 + x^3 + x^4 + x^5 + x^6 + x^7$

7. $1\cdot00995$

8. $1\cdot00504$

18a. 1. $-a.\sin ax$ 2. $-\dfrac{\pi}{180}\sin x^\circ$

3. $2\cos 2x$ 4. $-6\sin 3x$

5. $2x.\cos(x^2)$ 6. $-a.\sin(ax+b)$

7. $(2ax+b).\cos(ax^2+bx)$

8. $2\sin x.\cos x$, i.e. $\sin 2x$

9. $-3\sin x.\cos^2 x$

18b. 1. (a) $-\cot x.\operatorname{cosec} x$; (b) $-\sec^2 x.\cot^2 x$, or $-\operatorname{cosec}^2 x$

2. $a.\sec^2 ax$

3. $-2bx\operatorname{cosec}^2(bx^2)$

4. $\sin 2x + 2x.\cos 2x$

5. $\sec x(\tan x + \sec x)$

6. (a) $2\tan x.\sec^2 x$; (b) $2\tan x.\sec^2 x$

7. $3\tan x.\sec x(1+\sec x)^2$

8. $2\sec x.\operatorname{cosec} x(\sec^2 x - \operatorname{cosec}^2 x)$, or $\dfrac{2(\sin^2 x - \cos^2 x)}{\sin^3 x.\cos^3 x}$, etc.

9. $2(\sec^2 x.\tan x - \operatorname{cosec}^2 x.\cot x)$, or $\dfrac{2(\sin^2 x - \cos^2 x)}{\sin^3 x.\cos^3 x}$, etc

10. 1, max.; -1, min.

18c. 1. $\dfrac{2}{(1-x)^2}$ 2. $\dfrac{3x^2-4x-3}{(1+x^2)^2}$

3. $\dfrac{3(1-x)(1+x)^3}{(1+x^3)^2}$ 4. $\dfrac{4(1+x)}{(1-x)^3}$

5. $\dfrac{2x.\tan x - x^2.\sec^2 x}{\tan^2 x}$, or $x(2\cot x - x.\operatorname{cosec}^2 x)$, etc.

6. $\dfrac{x.\cos x - 3\sin x}{x^4}$ 7. $\sec^2 x$

8. $\dfrac{1}{\sec x} = \cos x$ 9. $\dfrac{5-42x-15x^2}{(3x^2+1)^2}$

10. $\dfrac{3\sqrt{x}-\sqrt{x^3}}{2(1-x)^2}$ 11. $\dfrac{1}{(1-x)^{3/2}.(1+x)^{1/2}}$

12. -6, max.; 2, min.

18d. 1. $\dfrac{-\sin x}{2y+\cos y}$ 2. $\dfrac{1}{y.\cos y+\sin y}$

3. $\dfrac{x-y}{x+y}$ 4. $\dfrac{4+y}{3-x}$

5. $\dfrac{\sin y+2x}{1-x.\cos y}$ 6. $\dfrac{2xy^3}{\sec^2 y-3x^2y^2}$

7. $\dfrac{2y^4}{x^3.\sin y+4xy^3}$ 8. $\frac{5}{12}, \frac{4}{3}$

19a. 1. $x = 1, y = 2, z = 3$

2. $p = 2, q = -1, r = 2$

3. $x = 3, y = 4, z = -2$

4. $p = 1, q = 1, r = 1$

5. $l = 3, m = -3, n = 0$

6. $x = -6, y = 3, z = 3$

19b. 1. $x = 1, y = 1$ and $x = \frac{45}{31}, y = -\frac{11}{31}$

2. $x = 2, y = -1$ and $x = \frac{2}{11}, y = \frac{39}{11}$

3. $x = -2, y = \frac{1}{2}$ and $x = \frac{18}{13}, y = \frac{1}{13}$

4. $x = 0, y = 1$ and $x = -\frac{172}{205}, y = -\frac{45}{41}$

5. $x = 1, y = -5$ and $x = \frac{41}{10}, y = \frac{16}{3}$

6. $x = -1, y = 3$ and $x = -\frac{25}{11}, y = -\frac{9}{11}$

19c. 1. $x = 1.56$, or -2.56

2. $x = 3.56$, or -0.56

3. $x = -1.5$, or -1.0

4. $x = 0.55$, or -1.22

5. $x = 0.63$, or -2.38

6. $x = 0.87$, or -1.54

7. $x^2+3x-4 = 0$ 8. $x^2+5x+6 = 0$

9. $x^2-5x+6 = 0$ 10. $x^2-x-6 = 0$

11. $8x^2-10x+3 = 0$ 12. $8x^2-7x = 0$

13. Two unequal real rational roots.

14. Two unequal real irrational roots.

15. Two coincident real rational roots.

16. Two unreal roots.

17. Two unreal roots.

18. Two unequal real irrational roots.

19d.   1. $cx^2+bx+a = 0$

2. $ax^2-bx+c = 0$

3. (a) $\sqrt{(\alpha+\beta)^2-4\alpha\beta}$; (b) $\alpha\beta(\alpha+\beta)$

  (c) $\dfrac{(\alpha+\beta)\{(\alpha+\beta)^2-3\alpha\beta\}}{(\alpha\beta)^3}$

4. $rpx^2+(2rp-q^2)x+rp = 0$

5. $4q^2 = 25rp$

6. $l^4x^2-(lm^2n-2l^2n^2)x+n^4 = 0$

7. $x^2-15x+25 = 0$

8. $x^2+8 = 0$

20a.   1. $-\dfrac{180}{\pi}\cos x° +c$ 2. $-\tfrac{1}{4}\cos 4\theta +c$

3. $\dfrac{180}{\pi}.\sin x° +c$ 4. $\tfrac{1}{5}\sin 5x +c$

5. $1-\dfrac{\sqrt{2}}{2}$, or 0.293 6. $\dfrac{\sqrt{2}}{2}$, or 0.707

7. $\dfrac{\pi}{3}+\dfrac{\sqrt{3}}{2}$, or 1.91 8. 1.5

9. 2π cubic units   10. π cubic units

20b.   1. 36 m   2. 96 m/s; 144 m

3. 6 m/s; 6.67 m   4. 12 m

5. 216 m/s²; 322 m   6. 486 m

20c. 1. $\frac{1}{6}.(x^2+3)^3+c$ 2. $\frac{3}{32}$

3. $-\dfrac{1}{4(x^4+2x)^2}+c$ 4. $\frac{3}{7}$

5. $3\cdot45$ 6. $2\frac{2}{3}$

21a. 1. $2y-5 = 0$
2. $5x-3y-1 = 0$
3. $x^2+y^2 = r^2$
4. $x^2+y^2+6x+4y+4 = 0$
5. $3x^2+3y^2+28x-14y+47 = 0$
6. $x^2+y^2-5x-4y = 0$
7. $4x-7y-29 = 0$
8. $y^2 = 4x$

21b. 1. $x^2+y^2-6x-6y+9 = 0$
2. $x^2+y^2-8x+2y+13 = 0$
3. $(0, 0)$, radius 5
4. $(2, 1)$, radius 4
5. $16x^2+16y^2-16x-24y-51 = 0$
6. $(-2, 3)$, radius 4
7. $x^2+y^2+4x+8y-5 = 0$
8. $(\frac{2}{3}, \frac{1}{3})$, radius $\sqrt{2}$

21c. 1. $x+y+2 = 0$
2. $5x+3y = 0$
3. $x+3y-10 = 0; x-3y-10 = 0$
4. $12x+5y = 0; 12x-5y = 0$
5. $x = -3; y = 3$
6. $4x^2+4y^2 = 37$

Part Two

22a. 1. $15\cdot4$ m   2. (i) $t = 4$ s (ii) meet after 5 s
(iii) 15 m from start

3. 10 and $14\cdot8$ s   4. $3\cdot33$ m/s^2

5. 18 m/s; 135 m   6. 1694 m

22b.   1. $\frac{10}{9}$ m/s^2; 225 m   2. 120 knots/hour; $2\frac{1}{2}$ n.m.
   3. 10 m/s; 30 m/s
   4. 1452 m; 22 s
   5. 31·6 m/s; 4·19 s
   6. 6·67 m/s^2; 40 m/s

22c.   1. 4·9; 19·6; 123 m   2. (a) 1960 mm/s
   2. (b) 29·4 m/s   2. (c) 141 km/h
   3. 323 m; 567 m   4. 1960 m; 40 s
   5. 10·2 s   6. (a) 16·6 m/s; (b) 6·7 m/s

22d.   1350 m   2. 5 s
   3. 24 m; 6 s   4. $\frac{48}{7}$ m/s^2; $2\frac{1}{2}$ s
   5. 983 m; $28\frac{1}{3}$ s   6. 0·5 m/s^2

23a.   1. 5·8 N at 20°   2. 6·1 N at 35°
   3. 5·0 N at 53°   4. 3·6 N at 74°
   5. 3 km/h at 56°   6. 18·5 m/s at 11°

23b.   1. 8·6 N at 53°   2. 5·4 N at 9°
   3. 3·5 N at 210°   4. 78 km/h at 18°
   5. 5·1 at 52°   6. 7·1 N at 340°

23c.   1. 5·82 N at 20° 6′
   2. 6·08 N at 34° 43′
   3. 5·00 N at 53° 8′
   4. 3·61 N at 73° 55′
   5. 3·01 km/h at 56° 15′
   6. 18·5 m/s at 10° 38′

23d.   1. 8·33 kg; 155 N   2. 0·3 kg
   3. 10·83 kg; 253 N   4. 221·5 N and 13·7 N
   5. 3·5 m; 882 N   6. 3·45 m; 4 m

24a.   1. Both 69·3 (or 49$\sqrt{2}$) N; 135°
   2. 2·04 N and 1·18 N; 0·75 m
   3. 0·57 N and 1·13 N
      Hung at B
   4. 136 N and 68 N; 1 m
   5. 17·8 N and 6·7 N; 0·36 m
   6. 0·89 N; 0·53 N; 0·93 N

24b.
1. $0.46g$ N and $0.20g$ N; 0.75
2. $5.45g\left(\text{or }\dfrac{27\sqrt{2}}{7}g\right)$ N
3. $38.2g$ (or $27\sqrt{2}g$) N
4. $1.35g$ N
5. (a) $3.52g$ N; (b) $2.31g\left(\text{or }\dfrac{4\sqrt{3}}{3}g\right)$ N
6. $45°$; $4.50g$ N

24c.
1. $0.35\left(\text{or }\dfrac{\sqrt{3}}{5}\right)$ 2. 0.13
3. 0.067 4. $15g$ N; $45°$
5. $0.29\left(\text{or }\dfrac{\sqrt{3}}{6}\right)$ 6. 0.17

25a.
1. 10.4 N at $155°$
2. 4.95 N at $134°$
3. 6.47 N; 8.13 N
4. 3.9 N at $37°$ with upward vertical; 2.9 N
5. Zero; 4.9 N vertically upwards.
6. 12.6 N at $50°$ to vertical.

25b.
1. Both 69 (or $49\sqrt{2}$) N; $135°$
2. 2.0 N and 1.2 N; 0.75 m
3. 0.6 N and 1.1 N
   Hung at B.
4. 140 N and 70 N; 1 m
5. 18 N and 7 N; 0.4 m
6. 85 N at $30°$ to upward wall; 49 N

25c.
1. $0.35\left(\text{or }\dfrac{\sqrt{3}}{5}\right)$ 2. 0.13
3. 0.067 4. 147 N; $45°$
5. 7.0 N at $98°$ to downward vertical
6. 48.6 N at $172°$ to downward vertical

26a.
1. 9.8 N   2. 10 kg
3. 0.029 N   4. 2400 m/s^2
5. 1.67 m/s^2 6. 10 000 N

26b. 1. (a) 490 N; (b) 550 N; (c) 490 N; (d) 430 N
2. 0·0196 N   3. 8·55 m/s^2
4. 1·8 m/s^2 downwards or retardation upwards.
5. 0·000 98 N; (a) 0·000 97 N; (b) 0·000 99 N
6. $m = 12·2$ kg, $f = 1·63$ m/s^2

26c. 1. 1·96 Ns   2. 45 N s
3. 2700 m/s; 5400 N
4. 38·3 m/s
5. (a) 5625 N; (b) 3750 N
6. 6·67 kg

26d. 1. 3·92 J
2. 337·5 J
3. 16 200 J; 7 290 000 J; 12 150 000 N
4. 93·3 J
5. 7670 N
6. 6·67 kg

26e. 1. 5400 W; 20 100 W 2. 432 kW; 31·1 km/h
3. 120 km/h   4. 0·28 m/s^2; 142 kW
5. 300 kW; 463 kW   6. 2400 N per 1000 kg; 1 in 5·5; 54 kW

26f. 1. 3·33 m/s; $\frac{1}{3}$ 2. 1·5 m/s; 9000 J; 3·6 m
3. 1·9 m/s; 3·9 s
4. (a) 3·6 m/s; $\frac{8}{35}$; (b) 1·2 m/s; $\frac{32}{35}$
5. (a) 0·575 m/s; (b) 0·175 m/s
6. (a) 67·5 m; (b) 90 m
7. 56 N; 4·54 m/s; 1·08 s
8. 2·56 m/s   9. 1·81 m/s

28b. 1. 50, 7   2. 49 (48$\frac{1}{2}$); 8 (7$\frac{3}{4}$)
3. 100; 28; 52; 24   4. 32; 36; 7; 2 (1$\frac{1}{2}$)

28c. 1. £970; £185; ±£10
2. £3·13; £0·21; ±£0·01
3. 10·1; 1·0; ±0·1
4. 59·70 cm; 0·80 cm; ±0·05 cm
5. 135; 26; ±1
6. 11·1p; 1·1p; ±0·1p

28d.   1. 93; 7; ±1
       2. 800; 225; ±25
       3. 62 m²; 38 m²; ±5 m²
       4. 57%; 13%; ±1%

28e.   1. £962; £187   2. £3·126; £0·217
       3. 10·07; 1·05   4. 59·70 cm; 0·835 cm
       5. 135·0; 26·25   6. 11·07p; 1·12p
       7. 92·1; 7·5   8. 807; 239
       9. 61·8 m²; 38·7 m²
     10. 57·6%; 12·5%

29a.   1. 15; 2·58; 2·22
       2. 21·8%; 29·7%
       3. 9·9; 5·79; 4·92; 4·90
       4. 1·08; 1·20; 1·19; 1·00
       5. 10; 2·83; 2·4; 2·4
       6. 20; 5·66; 4·8; 4·8

29b.   1. 22·28; 0·407
       2. 135·4; 4·67; 3·56
       3. 966·5; 21·75; 21·75; 25·96

29c.   1. 32·53; 2·10
       2. 64·13 cm; 1·72 cm
       3. 28·21; 2·24; 1·78; 1·77
       4. 5·55; 3·74
       5. 27·06p; 0·90p; 0·64p
       6. 5·889 m; 0·126 m; 0·100 m
       7. 0·389; 0·811

29d.   1. £977; £298; £237
       2. £3·136; £0·314; £0·245
       3. 9·89; 1·52; 1·25
       4. 59·70 cm; 1·31 cm; 1·06 cm
       5. 140·5; 35·8; 29·1
       6. 11·13p; 1·79p; 1·28p

30b.   1. 103·3   2. 123·8
       3. 107·7   4. 105·1
       5. 109·9

31a. 1. (i) $\frac{1}{8}$; (ii) $\frac{1}{26}$; (iii) $\frac{1}{104}$

2. (i) $\frac{1}{78}$; (ii) $\frac{17}{78}$; (iii) $\frac{2}{13}$

3. (i) $\frac{1}{36}$; (ii) $\frac{25}{36}$; (iii) $\frac{5}{18}$

4. (i) $\frac{1}{10}$; (ii) $\frac{9}{10}$

5. $\frac{2}{15}$ 6. $\frac{67}{132}$

7. $\frac{27}{55}$ 8. $\dfrac{1}{27,720}$

31b. 1. $\frac{1}{4}$; 105 2. $\frac{1}{2}$, 93

3. 32; 48; 24; 4 4. $\dfrac{761}{5^{10}}$

5. (i) $\dfrac{1}{390,625}$; (ii) $\dfrac{331,776}{390,625}$

6. $\frac{64}{81}$

31c. 1. $1; \dfrac{\sqrt{3}}{2}$ 2. $4; \sqrt{2}$

3. $1; \dfrac{\sqrt{6}}{3}$ 4. $8; \dfrac{2\sqrt{10}}{5}$

5. $\frac{4}{25}; \dfrac{4\sqrt{6}}{25}$ 6. $\frac{10}{3}; \dfrac{\sqrt{10}}{3}$

31d. 1. No evidence of bias

2. (*a*) no evidence; (*b*) significant at 5% level

4. 0·035, significant at 1% level

5. 7·0% to 15·4%

6. (*a*) none; (*b*) significant at 5% level; (*c*) significant at 1% level

31e. 1. 20·34 cm to 22·86 cm

2. 276·25 to 283·75

3. 272·5 to 287·5

4. 265 to 295

5. 72 cm ± 15 cm; 40 cm

6. 72 cm ± 15 cm; 80 cm

7. 48 cm ± 3 cm; 7 cm

8. 144

9. 5; 2·58; 0·861